T0296990

MACRO MARKETS

MACRO MARKETS

Creating Institutions for Managing Society's Largest Economic Risks

ROBERT J. SHILLER

OXFORD
UNIVERSITY PRESS

This book has been printed digitally and produced in a standard specification in order to ensure its continuing availability

OXFORD
UNIVERSITY PRESS

Great Clarendon Street, Oxford OX2 6DP

Oxford University Press is a department of the University of Oxford.
It furthers the University's objective of excellence in research, scholarship, and education by publishing worldwide in

Oxford New York

Auckland Cape Town Dar es Salaam Hong Kong Karachi
Kuala Lumpur Madrid Melbourne Mexico City Nairobi
New Delhi Shanghai Taipei Toronto

With offices in

Argentina Austria Brazil Chile Czech Republic France Greece
Guatemala Hungary Italy Japan South Korea Poland Portugal
Singapore Switzerland Thailand Turkey Ukraine Vietnam

Oxford is a registered trade mark of Oxford University Press
in the UK and in certain other countries

Published in the United States
by Oxford University Press Inc., New York

Reprinted 2007

ISBN 978-0-19-829418-4

Preface

This book arose from an awareness that the major economic risks that our society faces are dealt with individually: each person bears his or her own misfortunes. Society does share some risks, such as risks of natural disaster, medical emergencies, or temporary unemployment; society takes care of many sudden or extreme misfortunes. Still, most of the risk that individuals face about their lifetime well-being is not shared. We allow our standards of living to be determined substantially by a game of chance.

The sharing of economic risk is one of society's deepest concerns, and rightly so. Inequality of income and wealth is painful to see. Income and wealth determine who is served and who is servant, determine who may expect to live in comfort and health and who may not, determine who can pursue a fulfilling career or life plan and who cannot. To the extent that this inequality is created by pure luck, it is not only painful to see, but also a shame.

We must ask why our market system does not allow all people, not just the most extreme cases of economic hardship, to share their largest economic risks, and whether any new technology of markets, to be used by individuals or by organizations to which they belong or with which they have business, might make such risk sharing possible.

The initial spur to my thinking on this topic was living (in New Haven) through the sudden and spectacular residential real estate boom in the mid 1980s that afflicted the north-east United States. The prices of homes are an important part of the determinants of our standards of living, and at this time these prices sustained some major shocks. At the peak of this boom house prices were rising at the rate of about 40% per year; prices doubled in a few years. At this time it was easy to observe significant disruptions in people's lives. People who had not purchased homes realized that they might never be able to afford them. In the midst of this boom young people who did not feel ready to buy housing felt compelled to make great sacrifices to invest now in a home, or else risk being priced out of the market. Others (like myself) who had bought before the boom felt a little guilty to receive this windfall, which could be used now to borrow against or used later after selling and moving to another part of the country or to another kind of housing arrangement. What did I do to deserve this windfall? When had I asked to play this game of chance with my friends and neighbors?

Then real estate prices tumbled in the 1990s. Those young friends, who had put their savings in a house that they did not want yet, saw their savings wiped out, and found themselves unable to sell the house to pay off their debts. Lending institutions, who had provided mortgages to these people, were in trouble. Builders, who had rushed to meet the demand, found themselves in a difficult situation, with inventories of houses that sometimes could not be sold at cost. Why does society let this happen, I wondered? At least, why does society not create institutions that would help people deal better with such market dislocations? Or is this nonsense just part of living, which we cannot ever avoid?

Starting in 1990, Karl Case, an economist at Wellesley College, Allan Weiss, a former student of mine at Yale and a recent graduate of the Yale School of Management, and I launched a campaign to try to convince a futures exchange to start futures markets in residential real estate prices. A futures market, of course, can be described as a market for bets on the course of the price or index that defines that market; the primary purpose of these markets is not to enable people to gamble but to 'hedge', to make offsetting bets, to cancel out the bets that they have already found themselves making due to their economic circumstances. Someone who is a holder of real estate (that is, who is 'long' in the 'cash market') can in effect place a bet that real estate prices will go down (that is, can go 'short' in the futures market). Then, should real estate prices indeed fall, the winnings in the futures market would cancel out the losses in the real estate market.

To those familiar with hedging commodities or financial assets, it will be recognized that such risk-cancelling bets, in one form or another, are the essence of enlightened risk management. Hedging in such markets is essentially the same as buying insurance against price declines. Even though the contracts people sign in these hedging markets are not called insurance policies, such markets are essential to the provision of insurance, since they are the places where risks are pooled among the world's risk bearers. The services provided by real estate futures markets would be retailed to the general public by companies offering insurance policies or analogous retail products, who would themselves hedge in the real estate futures markets the risk they incur in providing these policies to the general public.

Karl Case and I got into the proposed real estate futures as an outgrowth of our joint work trying to describe and explain the real estate booms here and elsewhere, and Allan Weiss had been working on a housing equity participation plan that would mitigate the effects on homeowners of housing price movements. We reached the idea together that making the real estate market liquid by creating new international markets, such as futures markets, is the most fundamental step that could ever be taken to enable people and organizations to protect themselves from another such real estate fiasco. When they hedge (or deal with intermediaries that help them hedge) in such a market, they can in effect just announce that they are not playing this real estate game. We wrote articles, made presentations at futures exchanges and at financial and insurance institutions that might use the futures markets, and incorporated ourselves (as Case Shiller Weiss, Inc.). Allan Weiss is now president of this company; we are producing real estate price indices for possible cash settlement of real estate futures contracts. We are working with the Chicago Board of Trade to study the feasibility of certain futures markets in real estate prices.

This book resulted from my efforts to think through the issues involved in settling risk-management contracts in terms of real estate prices, and to generalize this thinking, beyond just the real estate application, indeed, beyond applications that will be easy to sell right away to designers of new contracts. The same kinds of contract that might be used to hedge real estate risk, the same principles of risk management, the same sorts of index number, the same kinds of retail institution, could be used to allow people to hedge all manner of risks to their standard of living.

Consider an example of what might happen if national income futures markets, and associated retail markets, became well developed. National income markets are one class (perhaps even the most important class) of new markets proposed in this book. National income futures markets would be futures markets for contracts that are settled on the basis of national income measures. Employers might buy national income insurance policies from insurance companies as a benefit for their employees, policies that insure these employees against downward shifts in the income level in their country. The insurance companies might then lay off in national income futures markets the risk they incurred by writing these policies. We might expect then that the risk of national income fluctuations would be borne by large, international institutional investors, who diversify by taking the other sides of these futures contracts in many countries around the world.

Suppose then that, for example, a change in economic conditions affects national income in one country (let us say Japan) adversely and in another country (let us say the United States) favorably. Individuals in Japan would not suffer; they would find that their own individual income was insulated from the income decline by their insurance policies. Insurance companies in Japan would also not suffer; they would find that their profits from their hedges in the Japanese national income futures market would offset their losses in payouts to Japanese policy-holders. Insurance companies that had insured American workers against income declines would face losses in the national income futures markets, and would be forced to pay out from their reserves to settle their contracts. Ultimately, there would be a flow of financial assets from the United States to Japan that would make possible the flow of real resources that would maintain Japanese living standards. The flows would, of course, go the other way, from Japan to the United States, if the fortunes of the two countries were reversed. That is the nature of risk sharing; *ex ante* neither country knows which way economic fortunes will turn.

These risk-sharing arrangements, as exemplified by the story of risk sharing between the United States and Japan, are not made by governments but by individuals and firms, and presumably any of these who want to opt out of any risk sharing can do so. The governments of the United States and Japan need not agree on any risk sharing, only to agree to allow the free market institutions of such futures or analogous hedging markets and insurance policies to exist. It would seem highly unlikely, from political considerations; that the governments of the United States and Japan would agree to such risk sharing for their citizens; fortunately there is no need for them to do so. Risk sharing could begin slowly, growing through time as more individuals, firms, and institutions realize its importance.

I will argue that we need a set of large international markets, markets for the major income factors shared by substantial groups of people: markets for national, regional and occupational incomes and for major service flows, such as real estate services. These markets may take the form of securities, futures, options, swaps, or other forward markets. A set of such markets would provide a choice of hedging vehicles that would enable retailers such as insurance companies to create policies that would allow people to protect their standards of living from important random shocks. In contrast, the risk management that is available to people with today's markets is minimal. Risks that are traded in today's financial markets are only a small fraction of our risks to standards of living. Consider, for example, the fact that our stock markets are claims only on corporate dividends and yet corporate dividends amount to only a few percent of national incomes, about 3% of national income in the United States.

Much of this book is technical, not easy reading: theory of market construction, of index numbers, and of statistics. The book is intended mostly for economists, for contract designers at futures and options exchanges, for originators of swaps and other financial deals, and for designers of retail products associated with risk management, such as insurance, pension plans, and mortgages. Yet I have tried to arrange the material so that a non-technical reader can follow most of the broad themes of this book. Until Chapter 6, most of the technical material is relegated to appendices that can be skipped without losing the thread of the argument; the concluding chapter, Chapter 9, contains no mathematics.

Acknowledgements

My colleagues Karl Case, Chuck Longfield, and Allan Weiss, by developing with me our proposals for new futures markets in real estate, and joining with me in promoting such markets in the United States, are the source of many ideas in this book. John Campbell, who was my co-author on a number of papers on speculative markets, was also a source of ideas.

There were many others who helped me in one way or another with this book, including Peter Abken, Donald Andrews, Peter Bernstein, Steven Bloom, William Brainard, Michael Brennan, John Clapp, Avinash Dixit, Peter Donnelly, Robert Engel, John Geanakoplos, Gordon Gemmill, Carmelo Giaccotto, William Goetzmann, Zvi Griliches, Sanford Grossman, Alan Heston, Jonathan Ingersoll, Ed Iversen, Theodore Keeler, Alvin Klevorick, Paul Kupiec, Larry Langowski, Bruce Lehmann, Hayne Leland, Ben Krause, Robert Miller, Evan Morton, Nathan Most, William Nordhaus, Todd Petzel, Steven Roberts, Richard Roll, Stephen Ross, Nouriel Roubini, Xavier Sala-i-Martin, Jeremy Siegel, Christopher Sims, Robert Summers, Steven Sural, James Tobin, and Niel Wilson.

The progress on this book was helped by my research assistants: Stefanos Athanasoulis, Kevin Burrows, Jose Carvalho, Margo Crawford, Vassil Konstantinov, Chiong-Long Kuo, Phillip Molner, Christopher Musto, Don Nakornthab, Chuin Hwei Ng, Todd Sandoz, David Waller, Toshiaki Watanabe, Murat Yulek, and Ramzi Zein.

I want also to thank the students at Yale who participated in my seminar 'The Establishment of New Markets', and helped me with their research and their comments on the manuscript for this book.

The research for this book was supported by the US National Science Foundation under grant SES-91-22229. Part of the research for this book was undertaken while the author was supported by a Guggenheim Fellowship. Research support was also provided by the Russell B. Sage Foundation.

Andrew Schuller and Anna Zaranko of Oxford University Press have been very helpful in bringing this manuscript to completion.

My gracious hosts at Oxford University during my Clarendon Lectures in May, 1992, where I presented an early draft of this book, offered thoughtful comments.

The staff and colleagues at the Cowles Foundation for Research in Economics at Yale University have been very helpful; Lois Jason and Glena Ames were particularly so. Shirley Kessel provided the subject index.

My wife Ginny gave unsparingly of her time and effort to support my writing, and she has been, as always, a sounding board and source of new ideas for my work. Finally, my two sons helped me with their patience while I put in a great deal of time on this book: my son Ben, and my younger son Derek, who also helped by informing people that his Daddy was writing a book on mackerel markets.

Contents

O Fortuna,
velut luna
statu variabilis,
semper crescis
aut decrescis;
vita detestabilis
nunc obdurat
et tunc curat
ludo mentis aciem,
egestatem,
potestatem
dissolvit ut glaciem.

from Carmina Burana

1
Introduction

Living standards are not fully insurable because of moral-hazard problems: if people or organizations knew that their incomes were guaranteed regardless of the amount of effort that they put in, then there would be a markedly reduced incentive to make efforts to maintain income. The effect of this moral hazard has been discovered in many times and places throughout history. It was discovered when, with the Speenhamland Law of 1795 in Britain, reformers raised workers' income to a specified level, the difference coming out of public funds. It was discovered again when idealistic socialists under Robert Owen set up communities in the nineteenth century where incomes were pooled. It was discovered again when communists in many countries in the twentieth century made attempts to move society toward the Marxian ideal of distribution, 'From each according to his abilities, to each according to his needs'. The humanitarian, idealistic, and revolutionary impulses that gave rise to these various social movements came up against a hard reality, the ill effects on incentives; most of these social movements were later abandoned. Moral hazard is not total: people and organizations do continue to function even if their incomes do not depend on it; but the inefficiencies created by such total insurance are so significant that they cannot be ignored.

Changes in living standards that are due to objective and quantifiable causes beyond an individual's or organization's control can, however, be covered by insurance policies without exposing the insurance companies to moral hazard from their policyholders. Thus, for example, the random decline in income (in the form of a decline in rents or real estate services) that a person faces because a house burns down is objective and easily measured; the risk of such decline is insurable. The moral hazard that the owner will deliberately burn down the house is small. In practice the objective fact of a fire is mostly not under the control of the homeowner. The decline in income that is caused by disability is similarly insurable; disability due to accident or illness can be objectively measured; most people will not disable themselves to collect on insurance. The decline in disposable income that is caused by medical expenditures due to health problems is also similarly insurable, as is the decline in income that is caused by the death of a family member.

Insurance policies on these objective and verifiable risks of income fluctuations have long been offered by our insurance companies to the advantage of the public. But the insurance industry has not devised policies that insure against many other causes of fluctuations in incomes. It is far more likely that a property will lose economic value owing to changes in economic conditions than that it will burn down. It is far more likely that an individual will face adverse labor-income shifts because of changes in the market for that person's labor than that that person will suffer a physical disability.

These economic causes of changes in standards of living that should be insurable without moral hazard because they are beyond individual control are still not insurable today because they are not so objective or easy to verify as fires or disabilities. The changes in the outlook for income, and the causes of these changes, are hard to describe. Economic changes may affect income only with long lags; the changes may become apparent only gradually through time, rather than catastrophically as with a fire. Insurance companies cannot verify whether an individual seeking income insurance is doing so because of private information that his or her own future income is likely to decrease; there is no medical exam that will verify that there is no pre-existing condition likely to lead to lower future income. This lack of objective evidence about the outlook for future income creates a problem for writers of insurance policies. But if we could create liquid markets for claims on aggregate income itself, then we could objectify the intangible economic causes of changes in aggregate standards of living, thereby making it possible to insure against adverse changes.

Individuals and organizations could hedge or insure themselves against risks to their standards of living if an array of risk markets—let us call them macro markets—could be established. These would be large international markets, securities, futures, options, swaps or analogous markets, for claims on major components of incomes (including service flows) shared by very many people or organizations. The settlements in these markets could be based on income aggregates, such as national income or components thereof, such as occupational incomes, or prices that value income flows, such as regional real estate prices, which are prices of claims on real estate service flows. Since the individual or organization has virtually no control over these aggregates, there is no moral hazard created by insuring the risks of these incomes.

If, or course, virtually everyone in a country were hedged against aggregate income risks, then there would be a problem of governmental moral hazard. A government, for example, could cheat by passing a law giving everyone a long vacation, thereby manipulating the national income. Such governmental moral hazard is not likely to be a factor until many people are hedged; even then there can be explicit contract provisions ruling out macro market settlements due to such government behavior. The behavior of governments can be monitored much better than that of individuals. A potential problem with governmental moral hazard is present even in existing financial markets, and has not prevented these markets from functioning.

Markets in claims on large income aggregates could be very inexpensive to trade in, could be very liquid. Just as there is less of a problem with moral hazard in these markets than in markets for individual incomes, there is less of a concern for dealers in these markets about being picked off by others with inside or superior information. There is simply less chance that others will have such superior information about aggregate incomes than about individual incomes. The modern information theory of transactions costs, Akerlof (1970), Copeland and Galai (1983), Gammill and Perold (1989), and Gorton and Pennacci (1991), suggests that transactions costs could be very low in macro markets. That transactions costs could be very low in these markets is of fundamental importance. Transactions costs are critical elements of the reasons for the forms of our existing financial markets and contracts; see Williamson (1979). Putting the same point in another way, the macro markets ought to lower the cost of information; information costs are a major barrier to the provision many kinds of insurance; see Kihlstrom and Pauly (1971).

It is odd that there appear to have been no practical proposals for establishing a set of markets to hedge the biggest risks to standards of living. Theoretical economists, ever since the contributions of Arrow (1964; 1974) and Debreu (1959), have appreciated the importance of the problem of incomplete markets, but the discussion of this problem seems to have remained at an abstract level. There have been some innovative efforts to expand the scope of markets, an effort, for example, to create real estate futures. But there seems not to have been any public recognition that we could really establish a group of markets to allow hedging of the major factors affecting aggregate incomes.[1]

Efforts to reduce the inequality of incomes might substantially be furthered by creating such markets to allow insurance against the major risks to income. Such efforts do not involve such politically difficult measures as taxing the rich and subsidizing the poor; it is in everyone's interest to insure themselves against income risks.

The revolutionary social thinkers described above could not see the fulfillment of their vision for communal sharing of income uncompromised by moral hazard unless there is development of a new community spirit, a new concern among the general public for others. But efforts to create such community spirit in society at large have not been successful enough to make possible the envisioned voluntary sharing. There are some stories of successes of utopian communes that completely pool all incomes, such as the Hutterites of North America, the kibbutzim in Israel, and the Itto-En and Yamagishi-Kai of Japan, but these communes, based on a sense of group feeling and intimacy built from shared experiences, are effectively large families. Usually a commune has no more than a few hundred members. Each commune is too small to allow much of the risk sharing envisioned here. Moreover, many of these utopian communes find that their community spirit declines somewhat through time and that private property becomes more important. Established communes may, for example, allow new members to retain possession of their preexisting wealth.

We should salvage as much of the objectives of these utopian thinkers as can be really achieved on a voluntary and self-interested basis for all of human society. People ought to freely share that component of their incomes still unknown and still to be dictated by forces beyond their control. And there is no other component of income but that which is still uncertain and beyond their control that self-interested people will voluntarily share. If we are to try to introduce as much communal sharing of income as can be achieved on a basis that is completely voluntary at all times then we will have to be, odd as this may seem, merely creating hedging markets for incomes.

This book, in proposing macro markets, describes a substantial array of new markets, and attempts to indicate ways around some of the major barriers to the establishment and success of these markets. The barriers are challenging, and require the adoption of some new technology, but they do not appear to be insurmountable.

An important barrier to establishment of these new markets is the difficulty in measuring inputs to the settlement formula for the contracts traded in the new markets. There are, of course, many statistics produced by governments and private agencies that might be the bases of settlement of risk-management contracts, but these measures were almost never created with the idea of using them for settlement of contracts. It is a very serious business to devise such measures, for if they are really to enable large numbers of people to manage their income risks, then truly vast quantities of money would have to change hands in response to the measures. To this end, some contributions to the theory of index numbers will be made in this book. Index numbers that are based on repeated observations of the same individuals or assets will be described, so as to reduce the problem of time inconsistency of the sample, a problem that seriously compromises many existing indices. It will also be argued here that the best way to generate an index of shocks to the stream of income may be to set up a new kind of market, a perpetual claim on index values, or an index perpetual futures market, that prices claims on long series of income index values. Such a market would then make use of public incentives for risk sharing to cause people to pool their information into a market price, which we can then regard as a more objective index number of the value of the income stream than we could construct directly from the income data. Prices generated in the perpetual claims or perpetual futures market could then be the basis of settlement of a wide variety of contracts.

Because of psychological barriers to individuals' participating directly in these markets, their reluctance to purchase some existing risk-management services, it is important that the market structure should be set up taking into account the behavioral traits that may limit participation. The nature of these psychological barriers will be studied here, and the experience from some past failures to set up insurance and hedging markets examined. Retail institutions will be proposed here to facilitate individuals' participation indirectly.

These important barriers to the establishment of the macro markets will be the subject of much of this book. First, however, let us get an indication of where we are heading, what kinds of markets we might hope to see established, and a broad sense whether such markets are within the realm of possibility.

The ideal: A world market for major income risks

The macro markets proposed here, perpetual claims, perpetual futures, options, swaps and analogous markets, could reduce the impact of changes in standards of living by arranging, in effect, to make it possible without excessive transactions costs, to share risks of such changes around the world. The losers in these markets (those individuals or organizations who shorted the markets in factors contributing to their own incomes and saw the value of their claims on these incomes increase) give wealth to the winners in these markets (who hedged and saw the value decrease) to compensate the latter for their reduced circumstances.[2] Those who decide to short their own macro markets who also go long in world macro markets are effectively deciding just to share income risks, the short-run movements in the market price in this period revealing the benefits or losses caused by agreeing to participate in such risk sharing during this period. Insurance companies could help people and organizations hedge by selling them insurance policies on their incomes and then themselves sell short in the macro markets, to hedge the risk that they incurred in writing the policies. The hedging markets may also function indirectly to reduce risks to standards of living, by encouraging firms to locate their production where incomes become low, by making it possible for the firms to hedge the risks of changes in these incomes.

Because moral hazard will be a very serious problem in insurance contracts for major components of income (whether individual income or organizational income), there must be strong incentives for those who sign such contracts to maintain their level of effort. One way that insurance companies might deal with the moral-hazard problem for income is to provide only partial insurance, that is, to sell policies only with large deductibles. Another way that insurance companies might deal with the moral-hazard problem is to write policies not on the individual's income or organization's income but instead on a macro factor that influences these. Those who are short in these contracts would collect if the present value of the index representing that factor declined. The risk that an insurance company incurs by writing such policies could be entirely laid off in a macro market.

In practice, the optimal way to write such policies may be to combine both of these methods. Those who retail insurance policies should try to write policies that insure against as much of the variation in an individual's or organization's income as possible without incurring undue costs due to moral hazard. To do this optimally would mean that the insurance companies could provide policies that insure fully (or nearly fully) against the declines in macro factors affecting that income, and perhaps partly (with a substantial deductible) against declines in income not due to these factors.

To do this, the insurance companies would undertake studies estimating the exposure of people and organizations to those income factors that are traded in the macro markets, and use the results of these studies to hedge for them optimally with a portfolio of macro market positions designed for their risks. Since there is no objective way to measure the news about people's expected future incomes, payouts on these policies would not be contingent on objective news like fires or illnesses, but on movements in the price of the portfolio. The partial insurance against declines in income not due to these factors would have to be predicated on some objective information besides prices in macro markets, and there will probably be difficulty defining such information. The partial insurance against declines in income not due to aggregate factors would be easiest to define where a cash market for the asset exists, as with residential real estate.

It would be natural for retailers to build the new risk management into existing products. For example, pension funds might alter their business to include some hedging of income risks. Or, life or health insurance policies could be amended to include some hedging of the policyholder's personal income risk. Employer contributions to pension funds could be debited and credited representing the employee's gains or losses in macro markets.

Homeowners' insurance policies could be designed to insure the service flow policyholders enjoy in their home against risks such as deteriorating neighborhoods or declining access to local jobs, and against possible loss in economic value of this service flow, so that when they move to another area or retire they can maintain this service flow. Alternatively, protection of standards of living could be enhanced by home mortgages that have their down-payment insured against loss due to declining sales price.

Labor unions might use these markets to help protect their members against adverse labor market conditions. In practice, existing labor contracts have aspects of risk sharing in them, and the contracts may allow the transference of the risks that individual workers face to the financial markets, by loading the risk of individual wage fluctuations in a firm into the price of that firm, a firm which is then held in diversified portfolios. But most labor contracts today run from one to three years. This horizon would seem to be extremely short, bordering on the irrelevant. Both workers and firms are concerned with their long-run interests; individuals look ahead to their retirement. These contracts are, therefore, not effective in reducing long-run risks, and are better described as temporary truces in the continuing market power war between labor and management.

One reason that labor contracts are so short may be that there is fluidity in the arrangement of the signers of the contracts. Individual workers often change jobs; employers find their own situation changing rapidly. Labor unions have been generally unable to create a cartel that monopolizes an entire labor market, and so are unable to represent individual workers who are in that market.

The role that labor unions might appropriately play in macro markets is just as another intermediary, one which might facilitate participation by their members in risk-hedging activities. Neither a firm nor a union representing employees at that firm would generally want to insure the individual by locking in a stable income at that firm. While such a guarantee by firms may be made more feasible by the possibility of hedging in macro markets, it is probably not advisable for a firm to make the guarantee in this form, since such a guarantee would lock the individual into the job at that firm. Rather, the firm should insure the individual against the effects of new adverse information about his or her lifetime earnings; doing this means hedging that individual in the appropriate macro markets, and passing capital gains and losses onto that individual, which that individual could then invest in a portfolio of claims on other income flows, such as world income.

Thus, income insurance policies will look a little like speculative market instruments, paying to the beneficiary different amounts at different times. People whose objective is to smooth their incomes will find that their income insurance policy gives them some very unsmooth payments. This will take some getting used to on the part of beneficiaries, who must be made to realize that the variation in payments is a natural characteristic of policies that insure them against drops in future income flows.

Hedging income risk in today's markets

In contrast to the ideal described above, there has been remarkably little attention paid to developing new methods for efficiently sharing risk about standards of living. All the discussion in theoretical finance about optimal diversification should, it would seem, have led researchers to an important mission: helping people diversify their major economic risks. What we have instead are little patches here and there, without any recognition of the ultimate objective of allowing full as diversification as possible of risks that are most important to individuals.

Taxes and transfers in place today represent partial insurance against income fluctuations. With an income tax, tax payments fall when individual income falls. With the unemployment insurance, welfare, and other federal programs, transfers will increase for an individual when individual income declines. These government programs, since they are enforced on everyone, solve the problem of selection bias that might plague private companies that tried to start income insurance plans (people who signed up for the plans would be those who have reason to expect that their own income was insecure); but the government programs do not solve the moral-hazard problem, the disincentive to work hard or at all that is created by the insurance.[3]

The federal tax system in the United States involves only a small amount of inter-regional sharing of risk. Sala-i-Martin and Sachs (1992) estimate, using data on the states of the United States, that a one-dollar reduction in a state's per capita income causes a decline in federal taxes of about 34 cents and an increase in federal transfers of about 6 cents. These figures were computed using a cross-sectional regression, and hence hold aggregate income constant. The federal government is, of course, unable to insure against aggregate income fluctuations without entering into risk-sharing agreements with foreign countries. What evidence there is suggests that inter-country risk sharing is negligible even within an organized common market. Sala-i-Martin and Sachs estimated that a one-dollar shock to aggregate regional gross national product within the European Economic Community (EEC) reduces tax payments to the EEC by only half a cent. Apparently it is difficult, politically, for countries to agree on risk sharing.

The kind of risk sharing that is imposed by income taxes and transfers is not optimal. The tax component represents incomplete risk sharing, and the transfer component of the risk sharing is limited to extreme cases.

Other forms of income risk sharing have been discussed. Friedman (1962) proposed that the current system of educational loans could be replaced by a system in which individuals trade shares in their future earnings for education; President Clinton has proposed an analogous plan, an income-contingent loan (ICL) plan (see Krueger and Bowen, 1993). Yale University in 1970 did in fact create a Tuition Postponement Option that involved such lifetime income risk sharing, Tobin (1973). The loan markets created by these programs do help with a clear problem in financing higher education (the inability to borrow against uncertain future income) but do relatively little for the bigger problem of sharing income risk.

Markets as inventions

We can draw inspiration in an effort to start new markets from the many inventions that have given rise to new kinds of markets, and that have changed forever the financial arrangements that we see every day. The history of economic institutions is one of punctuated equilibrium, where basic economic institutions remain largely unchanged for long periods of time, only to be superseded by new institutions whose advent can only be attributed to invention.

The very market system in which we now live was an invention, a system created by social thinkers who designed it with a purpose. According to a survey of research on primitive societies by Polanyi (1944), there is relatively little exchange of goods in these societies, and the exchange that does occur tends to be an external affair between communities, rather than among individuals within the community. Reciprocal gift giving tends to be more important than exchange, and hence there are few well-defined prices. Historically, laws to promote markets came as the inventions of social thinkers. These laws freed people from restrictive obligations to lord, clan, or extended family, and encouraged them to provide their goods and services to an impersonal market.

The proliferation of stock markets could not exist until the corporate law was created that defined the rights of shareholders and provided for limited liability of shareholders. Even the law defining corporations was not enough by itself to pave the way for the very liquid and large stock markets that we have in the United States, the United Kingdom and Japan. Other countries have had as many years' experience with stock markets, and yet there is relatively little activity in their stock markets. Bhide (1991) has argued that large, liquid stock markets are rather difficult to start, since agency problems of doing business are more naturally handled in firms each owned by a small number of investors with a long-term relationship with the firm, and that the liquid stock markets in the United States and similar countries are found only because of a conscious public policy designed to promote such liquidity.[4]

Futures markets could not be constructed until the idea came, apparently in the mid-nineteenth century, that a commodities exchange could guarantee performance of trades by requiring margin accounts and debiting and crediting these with daily resettlement of contracts, and that a commodities exchange could hire arbiters to make disinterested judgments about the quality of commodities delivered. This invention enabled price discovery for a standard commodity for the first time; it created liquidity.

The past couple of decades provide many examples of inventions that made new markets possible. The very idea of a financial futures market required some inventive activity; this bore fruit in 1972 when the Chicago Mercantile Exchange (CME) began trading currency futures. The idea of an exchange for options was a sort of invention which was first applied in 1973 when the Chicago

Board Options Exchange (CBOE) was established. The idea of cash-settled contracts was another invention, whose value was dependent on the proper specification of the settlement procedures. In 1981 the first cash-settled futures contract was created, the Eurodollar futures at the CME. Cash settlement was an important innovation, since it made possible the trading of contracts on theoretical index numbers, rather than solely on deliverable commodities. In 1982 the first index futures market was created, with the Value Line Index futures contract at the Kansas City Board of Trade.

These inventions have since been applied around the world. The arrival of these innovations cannot be attributed primarily to changes in economic conditions in the United States that fostered their application, since the markets were rapidly applied, after their first introduction, in diverse countries around the world. It would have been a mistake for anyone to try to explain in any economic terms why we did not have these markets a couple of decades ago; clearly the reason for their absence had to do with the fact that they were not invented (and proven) yet.

The functioning of market economies awaits new inventions. We can invent markets that allow the better handling of risks, and invent the associate retail institutions that allow individuals to make good use of them.

Markets as accidents of history

The development of markets has not proceeded at an even pace. Of course, all inventive activity proceeds by fits and starts. But the establishment of new markets is more uneven, since markets are social phenomena; the social, legal, and regulatory situation that fosters markets can inhibit their development, or even reverse and eliminate already existing markets.

It is in an important sense an accident of history that the ownership of profits of individual corporations and claims on obligations of governments is what is given most stress in our liquid private risk markets, rather than claims on economically more useful indices. This is an accident in the sense that the markets were created to allow entrepreneurs and governments to raise money, and to allow investors to cash in their investments when they will.

In fact, however, markets would better serve price discovery and hedging purposes if specific risks could be traded, rather than instruments whose price may have little meaning to the market at large. The information represented in the price of a share of a firm is possibly specific to that firm, and not readily generalizable so as to be useful to others.

Why is it that the risks that are traded in our large liquid markets are risks of individual firms, rather than risks associated with recognizable lines of business? Originally, the two were more nearly the same. Looking at a list of corporations in the United States from the turn of the century, one discovers that most of them appear to be one-product firms. However, the identification of firms with industry began to be blurred shortly thereafter. Notably, the merger movement of the 1920s, known for the tendency for vertical integration to occur, caused industries to merge with their suppliers or retailers. The merger movement of the 1960s, associated with the rise of large conglomerates, created large corporations each with a seemingly random variety of products. The result of these mergers is that the price discovery provided by our equity markets has no clear meaning in terms of risks of recognizable investments that others may be contemplating making.

It was an accident of history, of sorts, that stock index futures were able to be traded when they were in 1982; it was a 1981 agreement between the chairman of the US Commodity Futures Trading Commission (CFTC), Philip McBride Johnson, and US Securities and Exchange Commission (SEC) chairman John Shad that settled the jurisdictional disputes that had prevented the establishment of cash-settled financial futures in 1981 and stock index futures trading in 1982. Had the jurisdictional dispute ended up giving the SEC the control over these markets, their introduction might have been stalled.

It was an accident of history that financial futures markets began when they did: currency futures were not started until the dollar was freed from the Bretton Woods pegging of its value in 1971. We were fortunate that the regulatory environment in the United States would allow such experimentation. This first financial futures market was extremely important for the precedent that it set.

Accidents of history have played a large role in determining the kinds of futures markets that we now have, and the seeming failure of some innovative contracts was sometimes due just to extraneous factors.

Much has been made of the fact that the US experiment in establishing a consumer price index (CPI) futures market was a failure. The market, a futures market cash-settled in terms of the CPI, is a market that allows people to hedge inflation risk by making, in effect, offsetting bets on the course of the CPI, was proposed by Lovell and Vogel (1973). The idea received widespread acclaim; indeed, a futures market in a consumer price index would seem to be of major importance since it would permit converting nominal contracts (such as debt) into real contracts.[5] Despite the potentially revolutionary importance of such a market, its establishment was delayed, partly by regulatory delays, until 1985, when inflationary uncertainty had died down to virtually nothing. Perhaps low volatility of CPI futures could have sustained the contract had it already established liquidity, but the decline in volatility could not sustain a contract that was not already liquid (see Horrigan, 1987).

The CPI futures market had only a couple of flurries of activity, in 1985 and in early 1986. The market was launched June 28, 1985; 1,324 contracts were traded that year. By the end of 1985, however, the contract appeared moribund: on most days no contracts were traded at all. There was a sudden pick-up of volume on one day, January 21, 1986, when 189 contracts were traded. Volume continued to be between 100 and 300 contracts a day for most of the next month, and then volume gradually declined, reaching zero again in June 1986. The market never recovered: while 1986 volume was 8,776 contracts, 1987 volume was 2 contracts, 1988 volume zero contracts. What was it that happened on January 21, 1986 that spurred this flurry of activity, and why did the flurry end so soon? One can never be absolutely sure what caused people to trade, but it is noteworthy that on January 21, 1986, the news media ran headline stories that the price of oil fell below $20 per barrel, a fall of $2 in one day to the lowest price in six years. Over the next month, oil prices continued to plummet. The activity in the oil market may well have spurred interest in the CPI futures market, since the CPI is widely held to be strongly affected by oil prices. Over the month succeeding January 21, 1986 the implied inflation implicit in the CPI futures price steadily declined, reflecting the declining oil prices. It appears, therefore, that it was the increased uncertainty about inflation that prompted sudden interest in the CPI futures contract; the settling down of oil prices thereafter prevented enough further interest in the contract from developing. It indeed appears that the failure of the CPI futures contract was an accident of history, happening because the contract was launched in a period of stable prices, interrupted only briefly by this oil price shock.[6]

A futures market in the consumer price index did exist once in Brazil, and was a success there for several years. This futures market, started in 1987, was technically a futures market in the payout on government obligations, but since this payout was indexed to the monthly consumer price index, the market was in fact a market in the consumer price index itself.[7] The market was ultimately shut down by the Brazilian government, which feared that any form of indexation made the price level there more unstable. The closing of this market may be regarded as another accident of history, an accident caused by a popular theory that such contracts promoted inflation; had there been another government in charge, the contract might not have been terminated.

Following proposals by Miller (1989) and Gemmill (1990), futures markets in real estate were attempted by the London Futures and Options Exchange (London Fox) in 1991. There were both commercial and residential futures contracts.[8] The commercial contract was settled on an index of appraised value. The residential contract was settled on a hedonic price index derived from actual selling prices of individual homes. Unfortunately, the contracts were traded only for a brief time, from May through October. Trading volume had been very low, and the exchange had allegedly attempted to create the false impression of high trading volume by false trades.[9] Reportedly, only 7% of the reported trades were genuine. When the deception was reported, the market was closed; officers of the exchange resigned.

The London Fox was a small exchange without the resources to launch a major public education campaign, beyond a few seminars and mailings of brochures. It did not seek out and develop concrete uses of the market for its best potential clients. It certainly did not pave the way for the kind of new retail institutions that would use the futures markets; there were none of the retail risk-management institutions described above; no mortgages with down-payment insured against real estate price declines, no home value insurance.

The London Fox real estate contracts were launched at a time when UK real estate prices were falling slowly and steadily, and there was low turnover of real estate. This might be considered to have been an ideal time to launch such a market, since price declines are what motivate long hedgers. But this was a time of little excitement in the London real estate markets, with little action, and that fact may not have encouraged the development of much interest in this market.

Perhaps the biggest accident that befell the London Fox experiment were the alleged attempts by the management of the exchange to promote the contract through false trades, efforts that led to the premature demise of the contract. The contract was not available for trading long enough to give the market a good try. Given the inadequate public education campaign that accompanied its launching, people needed more time to get into this market, and they thought that they had time. Futures contracts sometimes do get off to a slow start, and highly-innovative contracts like this are all the more likely to need time to grow.

Evidence that accidents of history play a major role in determining the kinds of markets we have suggests that determined efforts to change our markets may succeed, not at the first try perhaps, but eventually. If we understand the role of accidents in developing our markets, we can move to make fundamental new changes in the markets, like those envisioned here. We have now to develop clearer plans for the kinds of markets we should have, taking a clear break from the past, and trying to define a system of markets that allows more rational risk management.

2

Psychological Barriers

One reaction to the proposals for new markets made here is that the public will not want to deal in them, because of some psychological inability to appreciate the benefits that the risk management may offer. Such a reaction is often given by people in the futures and options industry, who have seen many innovative contracts fail. It is often said that futures markets succeed only where there are professional dealers in the commodity or security traded who wish to hedge an inventory; for the national income and some other futures described here, there are no such dealers. The general public, it is asserted, will not trade directly in such futures, and if the reason they will not is that they do not appreciate the benefits of risk management, then it may be difficult for retailers of risk-management services to repackage the contracts in a form that will interest them.

However, the general public, including both individuals and firms, does use some risk-management services, notably insurance policies. The insurance industry has not failed to market risk-management policies to the general public; rather, their success has been highly variable.

Purchase of life insurance by individuals is quite widespread. According to the Life Insurance Fact Book, 81% of US households owned life insurance in 1984. Public appreciation of health insurance is also widespread. But purchase of some other kinds of insurance is less frequent. In 1984, only 22% of the working population carried long-term disability insurance (Cox *et al.*, 1991). There is no logical reason for people to skip disability insurance. The frequency of disabilities is much higher for the working population than the frequency of deaths, although most disabilities are eventually overcome. The economic loss occasioned by long-term disability is even larger and more catastrophic than the loss occasioned by death, since with long-term disability the family not only loses the income but is also burdened with costs of care for the disabled worker. However, the fear of disability has had less sales potential for insurance agents than fear of death.

Kunreuther (1977) has shown that the insurance industry has had great difficulty in selling such things as flood insurance to home-owners in floodplains or earthquake insurance to people on geological fault lines, even though some such policies made eminently good sense and were subsidized by the government. Most farmers do not hedge their crops in futures markets, or use other retail risk-management products for the selling price of their crops (see further, Chapter 5).

Do the proposed macro markets most resemble the life insurance policies for which there is a ready market? Or do they more resemble the disability or flood insurance markets or farmers' hedging markets? To try to answer this, let us turn to the psychological literature on the demand for insurance.

Research on psychology and risk perceptions

Research by decision theorists has, unfortunately, not resulted in accurate theories of how people make decisions about risk and insurance. The conventional theory of expected utility maximization, the standard paradigm for much of contemporary economic theory, would, subject to some qualifications concerning selection bias and moral hazard, imply that people would insure all risks. But there is no widely accepted alternative to the theory of expected utility maximization.

The most important alternative to the expected utility theory, at least for applied researchers in economics, has been the prospect theory of Kahneman and Tversky (1979). This theory, however, has a number of parameters that may vary from circumstance to circumstance, and has no clear implications for the proposed insurance mechanisms here.

One aspect of prospect theory is what Kahneman and Tversky called 'risk-seeking in the domain losses'. While people are risk averse when they are weighing alternatives involving only gains, they are supposed to seek out risk, in a sense, when weighing alternatives involving only losses. Kahneman and Tversky asserted this because of people's answers when asked to choose among various packages of losses. They asked their experimental subjects, for example, which they would prefer: a sure loss of $3,000 or an 80% chance of a loss of $4,000. Ninety-two per cent of their subjects preferred the 80% chance of a loss of $4,000, even though the expected loss was greater (at $3,200) in this case. (This is in contrast to their results with gains: when asked to choose between a sure gain of $3,000 and an 80% chance of a gain of $4,000, 80% of the subjects chose the sure gain.) This risk-seeking behavior in the domain of losses might then explain why people sometimes do not buy insurance policies. Unfortunately, the theory seems to do more than that: people's unwillingness to pay as an insurance premium the expected loss would seem to imply that no one could ever make a business of selling any kind of insurance.

But prospect theory does not clearly have this counterfactual implication, because of ambiguity in the theory as to what constitutes a loss.[1] An important determinant of behavior is what psychologists call framing, defined as the context or reference point suggested by the experiment or by other circumstances. If people were wealth framers, rather than the income framers described above, taking as a reference point zero wealth rather than their current wealth, then the description of the above choice involving losses could be viewed by the subjects as involving only gains: a choice between the subject's current total wealth minus $3,000 and, as the alternative, an 80% chance of current total wealth minus $4,000 and 20% of current total wealth. Then, since they are in the domain of gains where they are risk averse, they would be willing to pay enough for insurance to allow an insurance firm to make a profit (see Camerer and Kunreuther, 1989).

Since prospect theory does not include a theory of framing it would not appear reliable in allowing us to predict what kinds of insurance policy people will buy. Indeed, perhaps the most important single result from the literature on decision making under uncertainty is not theories like prospect theory, but the observation that framing is terribly important. Kahneman and Tversky and others found that they could vary substantially the outcomes of their questions just by the wording of the question, without changing the substantive choice people were asked to make. Even such apparently inessential details as what to call the losses or gains can make an important difference in the responses. Experimental studies have shown that people are more likely to choose a sure loss over a larger probable loss if the former is called an insurance premium (see Fischhoff *et al.*, 1980).

Framing can be changed by discussion; the experiments in the literature on demand for insurance did not allow subjects to consult with experts or to read the articles offered by consumer reports on the matter. Decisions whether to insure substantial fractions of one's income are deep and fundamental; presumably there could be substantial public discussion and hence more of a likelihood that individual behavior would satisfy higher standards of rationality than has been observed in experiments.

Lack of information and ambiguity

The experimental evidence cited above was produced in a situation in which people were told the exact relevant probabilities. In fact, however, people generally purchase insurance without knowledge of the relevant probabilities, and cannot verify whether they are overpaying or underpaying for insurance. This may explain, for example, why people apparently dramatically overpay for airline flight insurance (Eisner and Strotz, 1961). They may imagine the probability of crash to be much higher than it is.

The problem may not be just lack of knowledge of probabilities among potential purchasers of insurance; there may be fundamental lack of objective evidence about these probabilities, what may be called true uncertainty (Knight, 1921) or ambiguity (Ellsberg, 1961). The nature of human behavior in the face of such ambiguity is difficult to model.[2]

Einhorn and Hogarth (1986) postulated that people have an intuitive response to ambiguity that looks for, and over-reacts to, some analogy or suggestion of the probability; a suggestion that they call an anchor, from which they imagine other possible values of the probability, by considering other values in its vicinity. The outcome of this thought process is an intuitively revised probability; the outcome is dependent on the initial anchor which may vary greatly among individuals. Insurance may tend to be sold when insurance firms respond with lower assessed probabilities than their customers. Experimental insurance markets constructed by Camerer and Kunreuther (1989), involving subjects playing the roles both of individuals and of insurance firms, found that differences in anchoring could influence insurance prices and sales volume.

When there is such fundamental ambiguity about probabilities, other psychological tendencies may come into play in making decisions. The vividness of the potential loss may make a difference as much as any estimate of probability. Vividness may be influenced by past experiences, which may focus attention on the costs, or by circumstances or beliefs that serve to heighten awareness of the loss. Kunreuther (1977) found that the decision to purchase flood or earthquake insurance was not very much related to education or income levels, but that those who bought tended to be those who had personal experience with the disaster, even if that experience was obtained while living in another community where the probability of floods or earthquakes may have been completely different. The decision to purchase airline trip insurance may be more related to the imagination, and the vividness of the potential risk, than to any objective probabilities. On this account, it is hard to predict how vividly people may weigh the prospect of substantial decline in their income or the decline in their property values; this depends on times and circumstances.

Demand for insurance by firms

It might seem that we should be in a somewhat better situation to predict the demand for insurance by firms and other organizations than we are for individuals. Organizations tend to be run by professionals, who allocate time and care to decision making, and who might therefore be less vulnerable to inconsistencies caused by framing or by other psychological factors.

Much of the use of futures markets has in fact come from firms, corporations and partnerships, rather than from individuals. The reason for the importance of firms as buyers and sellers in these markets may be that only the kinds of professional who are commonly found in these organizations are able to understand and properly use these markets. The professionals have the time and incentives to take careful stock of these risks. Many of the apparently irrational departures from expected utility maximization that psychologists and economists have documented in individual behavior may be attributed just to costs of undertaking calculations correctly, given the limited cognitive capabilities of individuals and the limits on their time to do calculations (March, 1978; Slovic and Lichtenstein, 1983).

It may, on the other hand, seem odd that corporations have been so important in futures markets. Some economic models suggest that corporations would not rationally buy any insurance, or hedge any risks. The effect of the firm's risks is borne by shareholders and debt-holders of the corporation, who can diversify their risks in financial markets. According to this argument, risk that is idiosyncratic to the firm, the kind that can be insured, can also be diversified away by investors in existing stock markets just by holding a diversified portfolio of stocks.

The reason firms actually insure has to do with the 'firm-specific value of the team, its institutions, and its assets' (De Alessi, 1987). The value of each firm as it is presently constituted may be threatened by bankruptcy proceedings. The commitment of employees to their firm, their willingness to participate as a team and themselves invest in firm-specific human capital, is enhanced by their sense of relation to a going enterprise. Evidence that such firm-specific capital is important can be found in the fact that few firms are ever set up for a specific venture, to be dissolved shortly thereafter. The reason firms insure must be fundamentally connected to the reason firms endure.

The expertise that is available within firms may also help individuals make use of risk-management services for their own personal risks, since the decisions to do so may be made collectively among persons affiliated with a firm.

Social-psychological factors

Social psychologists have long known that difficult or ambiguous decisions tend to be influenced by social-group perceptions. The opinions of peers are demonstrably very important in individuals decisions, often more so than a logical argument received from experts or the media (see McGuire, 1969).

The human tendency to look to others when making difficult decisions may also have a rational component; people learn that others often know something, and that it may be wise to imitate others when in doubt. Rational equilibria can exist in which social movements begin because of a mistaken impression that others have a good reason for the movement. These social movements have been called informational cascades; see Bikhchandani *et al.* (1990) and Welch (1990).

Kunreuther (1977) found that, among the people he studied regarding flood and earthquake insurance, knowing someone else who had purchased insurance was strongly positively related to the decision to purchase. Katona (1975) found, from surveys in the 1950s and 1960s, that a strong predictor of individuals' decisions whether to own corporate stocks was whether they had any friends or relatives who owned them. Katona used the term 'social learning' to refer to the process whereby a large group of people exchange information and reinforce each others' interpretation of this information. The process of social learning takes time; whether it continues seems to be capricious; social attention may or may not be focused on these matters.

Social movements usually involve opinion leaders, influential people who make their opinions known to a larger group, providing some stimulus for social learning. This is why test marketing of a product on a small group is often unsuccessful in predicting demand for the product; the test marketing cannot simulate the kind of large-group interaction that will ultimately decide the fate of the product. Recall, for example, the experience of the test marketing of the new formula for Coca-Cola, which produced spuriously positive outcomes, and the marketing fiasco when the product was launched onto the entire market.

Because the process of social learning is rather hard to predict, there is no alternative to trying to launch the new markets nationally in a major drive, hoping that a social movement will take root that will produce a demand for the risk management made possible by these markets.

Gambling behavior

In trying to understand psychological barriers to proper hedging of risks, an important fact to consider is the tendency of some to court risks for no expected gain, to gamble. Because of the gambling impulse, we see institutions designed to create risk as well as institutions for hedging and insurance. Moreover, some people have the impulse to use hedging vehicles to create risk.

Interest in gambling is quite widespread, and in some people it reaches pathological proportions. According to a 1974 study (Kallick *et al.*, 1975) 61% of the adult population of the United States placed some kind of bet that year, and 1.1% of men and

0.5% of women are probable compulsive gamblers. Compulsive gambling is an addiction, which can lead to personal and financial ruin. Compulsive gamblers, and borderline compulsive gamblers, are probably attracted to speculative markets, and their behavior must have some impact on the markets. Psychologists have shown that an 'illusion of control' is a common human error, a tendency to believe in one's own judgment, a tendency to be optimistic about one's own luck, Weinstein (1989*a*; 1989*b*). This tendency can contribute to volatility in financial markets.

To the extent that gambling is a form of entertainment, and thrives only for certain kinds of risk for certain people, it is not itself likely to be a major barrier to the proper functioning of most risk-management products. But perhaps the impulse to gamble might be considered in a sense to be more important than a form of entertainment: to be instead behind the entrepreneurial spirit, to be part of the motivating spirit of the business world, and therefore to be behind much or most decision making. Even so, behind this same spirit is a desire to manage most risks, to play the game to win, not to see chance determine most outcomes. Research on gambling behavior has stressed that most gamblers have preferences for activities that offer them some sense of control and mastery, activities that support a feeling that they know well how to play the game, and this means that they do not seek to let chance alone dominate their outcomes. Thus, for example, highly speculative investors may be among the most avid users of risk-management products. Speculators motivated by gambling impulses may, for example, have a desire to take chances that a certain stock will do very well, and at the same time they may hedge the market risk of this stock by shorting the stock market.

The human tendency to be aroused and excited by risks that offer potentially large gains is probably a beneficial tendency itself, which has evolved in nature because it contributes to long run survival. The tendency may in some circumstances be seen to contribute to chaos and randomness in our lives; with financial markets it may be part of the genesis of speculative market volatility. But the basic impulse to take risks is so deeply ingrained, and such an integral part of our intellects, that it is very hard to know what policy might help alleviate the costs of this randomness without creating costs of a different nature.

It should be noted that there is a chance that the gambling impulse may sustain a speculative market, like a futures market, that is temporarily not interesting to hedgers, thereby promoting the survival of these markets. But the gambling impulse alone will probably not support a futures market for long. It is plausible that it will not, since if a futures market were to survive only on its appeal to gamblers, this market would have to compete with other forms of gambling that are designed to appeal to gamblers. Gambling behavior is not pure risk-taking behavior; gamblers do not just want to take risks; they also want to play a game that is attractive to them. Current laws prevent futures markets from attempting to be attractive to gamblers; so it would not seem surprising if gamblers are only incidental to such markets.

Speculative behavior

Speculative behavior is defined as risk-taking attempts to profit from subjective predictions of price movements; taking chances trying to buy low and sell high. Speculative behavior is not necessarily gambling behavior; speculators are not necessarily attracted by risk itself or by a sense of play. Speculative behavior normally helps assure that prices reflect underlying fundamental values. If, for example, futures prices for an agricultural commodity are unusually low at planting time, farmers might be discouraged from planting much of that crop. The consequence would be that not enough of that crop is produced, and so there would be low supplies of that crop and high prices when the crop is harvested. Speculators who recognize that the price of that commodity is temporarily low at planting time, and who are willing to take the risk of betting that the price will rise, may step into the futures market and buy contracts, thereby pushing the futures price up. The speculator thereby prevents the low supplies and high prices that would otherwise have been seen. The profits that a speculator earns in the market may be regarded as compensation for the work the speculator does in collecting information and providing it to the market.[3]

But speculative markets also appear to show a great deal of volatility which appears to be due to destabilizing speculative behavior. There may sometimes be, for example, an irrational bubble. An irrational speculative bubble is defined as a protracted price increase caused by a sort of vicious circle: when many people think that prices will go up, their contribution to demand may cause the price to rise further. Each round of price increase attracts more investor interest and hence generates yet another round of price increase. This sequence cannot go on forever: eventually the bubble must burst, and prices fall sharply; in this sense the bubble is irrational. In fact, this bubble story, which suggests that there would be serially correlated price increases followed by a sudden break, is very special, and in fact prices in speculative markets do not tend to show this time pattern, at least not very consistently. Nonetheless, while reality is more complicated, there appears to be an important element of truth to this story.[4]

The typical story of speculative bubbles involves people's reaction to price movements only. But, in fact, speculative price movements appear also to involve interpersonal communication and reaction of investors directly to each other as well; there are fashions and fads. Speculative price increases have occurred as well in real estate markets where there has been very imperfect, and lagged, information about prices. And, of course, all variety of social movements occur where there is no price at all to be observed. It is likely to be most accurate to think of speculative price changes as just another manifestation of the social behavior that is studied in broader contexts by social psychologists.

Such mass behavior, since it is generated by social movements, generates correlated risks—risks correlated across assets and hence not completely diversifiable. Insurance companies dealing with purely idiosyncratic risk, such as the risk of death of individuals or the risk of fire in homes, can eliminate any risk to the firm itself by selling policies that represent a very small part of the world economy. Selling a few thousand life insurance policies involves reducing almost all of the risk to the company, even though a few thousand policies represent a minuscule fraction of the world economy. Not so with risks that show substantial correlation with each other: an insurance company could deal with these only by making some risk-sharing arrangements, as by laying off the risks in a hedging market. A tendency towards destabilizing speculation may therefore be not so much a barrier to the establishment of macro markets as a reason to establish them. Creating liquid markets may thus make it more possible for people, by hedging, to insure themselves against the risk that they will be hurt by destabilizing speculation.

The apparent excess volatility in existing financial markets (LeRoy and Porter, 1981; Shiller, 1981; 1989) is likely to be substantially due to destabilizing speculative behavior. Other evidence consistent with the notion that destabilizing speculation is important is the tendency for loser stocks (stocks whose prices have declined greatly in the past few years) to have higher than average returns, for winners (whose prices have increased a lot in the past few years) to have lower than average returns (DeBondt and Thaler, 1985), for the variance of returns to increase less than proportionally with the time interval over which returns are measured (Poterba and Summers, 1988), and for stocks whose price is high relative to fundamentals to have low returns subsequently, and stocks whose price is low relative to fundamentals to have high returns (see e.g. Fama and French, 1988*a*; 1988*b*; 1992).

The existence of these speculation-induced price movements is a cause for concern about the establishment of major new hedging markets. The issue is not just whether the markets will succeed, but also whether the markets will be beneficial to society. The significance of destabilizing speculation is discussed further in Chapters 3 and 9 below.

Promoting proper public use of macro markets

The success of risk-management products for the general public, products issued by firms (such as insurance companies) who in turn hedge the acquired risk in macro markets, depends on proper public education in and proper design of the products. These retail products may be successful after a public information campaign acquaints the public with their value as hedging, rather than gambling, vehicles. Success in marketing retail risk-management products is made more likely if they are designed so that they serve clients well and are attractive to them.

Public education can be facilitated if retail firms offer their products in a social nexus, so that the social-psychological effects noted above may work to their advantage. When income insurance is offered as an employee benefit, the social situation of the workplace where such benefits are discussed may make for a far more appreciative audience than if an insurance salesman tried to sell such a policy directly to individuals. And when the employee-sponsored insurance policy is proposed by the firm or labor union as part of negotiations, the policy may in this context be carefully studied. In such a social nexus, more public attention may be given to the advantages of the risk management, there may be more inter-peer discussion of the advantages, and there may be opinion leaders that facilitate its acceptance.

Another strategy that retailers could use to make risk-management products more palatable is to combine them with other risk-management policies that have already won acceptance. In doing this, they drive home the analogy between the kind of risk management already commonly practiced and the new kinds of risk management, and allow people to view the new products as if they were already familiar. For example, as already suggested in the preceding chapter, home price insurance could be introduced as part of fire, flood, and other insurance policies on the same home. Home price insurance could be much more attractive if it were among the options that the homeowner could choose on the insurance application form. Decisions among insurance options are already made with the advice of professional insurance advisers, who would be at hand for the decision to purchase home price insurance as well. Or the mortgage insurance that is now associated with home mortgages could be replaced with a more comprehensive insurance that protects the down-payment against adverse real estate market turns; again, at mortgage application time the homeowner may well find it easy to choose to buy such insurance. Moreover, when the down-payment is thus insured against adverse home price movements, then some of the risk of default is reduced to the benefit of the mortgage lender; in this case the lender may also be able to make the policy more attractive by offering a lower mortgage interest rate.

Including new risk-management products as parts of other contracts may also contribute to acceptance because of an add-on effect: when someone is already making a decision to buy, other products that can be viewed as add-ons to the item being purchased may be readily purchased. Thaler (1991) has argued that, given the risk-seeking behavior for losses hypothesized by prospect theory, the theory of mental accounting implies that salespeople will try to segregate gains and integrate losses: this means describing separately to the customer the advantages of the add-on, but merging its price into the overall price.

According to theories of psychologists Daniel Kahneman and Amos Tversky (1981, p. 456), 'insurance should appear more attractive when it is presented as the elimination of risk rather than when it is described as a reduction of risk'. Thus, one way for retailers to tailor their product to individual's needs is to offer a policy that, in some sense that is meaningful according to the framing adopted by their clients, virtually eliminates some risk. Of course, no insurance can ever reduce all risks that people face; but since people show tendencies mentally to compartmentalize risk, it is possible, if such tendencies can be understood, that a policy could be designed that is widely viewed as eliminating a sort of risk, policies such as those (to be discussed in Chapter 5 below) against risk to the value of an entire farm rather than against risk in the price in one of the crops in one harvest of the farm. Individuals may find it attractive if they are offered insurance against adverse shifts in their wage rate in such a narrowly defined occupational and geographical area that the policy might be viewed as virtually eliminating risk.

Psychological barriers: summing up

Some of the psychological barriers discussed above may seem discouraging for our efforts to establish the most important new risk markets. The most discouraging fact is that, as noted at the beginning of this chapter, people do not routinely buy many risk-management products, and the range of commodities covered by futures markets remains narrow.

But we have stressed here that the kind of risk that is to be hedged in macro markets may be regarded as the most important kind of economic risk that people and many organizations face; and really just about everybody shares some exposure to macro risks. The kind of decision making that people and organizations might undertake for such risks can be very different from that which psychologists explore in their experiments, or that which individuals make when an insurance salesman tries to get their attention. The psychological theories that have been proposed to explain why many do not insure are not strong enough to tell us when they will or will not do so at times of major institutional change.

It is, of course, wise to entertain some doubts that the methods described here of dealing with psychological barriers to new risk-management products may ever work. But let us not be pessimistic. Public interest in any new market, as was noted above, takes the form of a social movement; such social movements are inherently hard to predict. They may be once-and-for-all things, like new inventions. It appears that no one has ever really tried to promote such a large social movement towards hedging most of the major risks to incomes. Since the hedging of such risks makes basic good sense, it is well that we work on the assumption that, when people give them enough attention and care, the psychological barriers to their use will fall away.

3

Mechanisms for Hedging Long Streams of Income

Most of people's income comes in long streams over many years; the most important insurance need that they face is guaranteeing this stream of income. We have already noted that conventional hedging and insurance mechanisms do not facilitate guarantees on the most important income flows. This chapter will propose instruments that make this possible.

That most people are interested in hedging risk of claims on income flows, rather than the risk of the next month's or next year's income, should be stressed. This means that any market that allows hedgers to protect themselves from information that is adverse to their income over the indefinite future must be a market in the present value (capital value) of this stream of income, not a conventional futures or options market in the income itself.[1]

Note that most futures and options markets today are constructed to price capital values, rather than to price claims on income at the maturity of the contract. There are many stock index futures and options markets, but nowhere in the world today is there any futures or options market that cash settles on the basis of the dividend accruing to a stock index portfolio at the final settlement date; exchanges would have found it very natural to create such markets had there been any interest in them. While we cannot rule out that there might be some people who would want trade in a conventional futures or options market cash-settled on the next value of a dividend index, it would appear that people are more concerned with hedging the value of their investments, which value collapses information about the indefinite future of the dividends accruing to these investments into today's price. By the same token, individuals are more interested in hedging the risk of their lifetime income than of their next paycheck, in hedging the risk of the value of their homes rather than the next rent payment.

As was indicated in Chapter 1, any insurance mechanism that is effective in providing such a guarantee is inherently a mechanism that creates or destroys value today in response to new information about future, even distant future, income flows. If there is a policy that promises to pay in the future any shortfall in the individual's income, then the existing policy will inevitably have sharply changing values as new information about the individual's own future income flows arrives. So long as the individual is not locked into this policy, then that individual is necessarily making a capital gain or loss in response to new information about future income flows. People will not want to be locked into an existing policy: they may wish to increase or decrease their insurance in accordance with changing circumstances, or to change the nature of their insurance; so they should ideally be able to sell their policies and purchase different policies. And such sales and purchases must reveal capital gains or losses.

Any insurance policy that guarantees an income flow against all such adverse events will, then, necessarily show apparently erratic changes in value from day to day, changes that probably cannot be attributed to any news whose meaning is objectively verifiable. There will be, inevitably, disagreements about the importance of any news, and the nature of the news that will come along cannot be foretold; this accounts for the difficulty in creating useful insurance policies against bad news *per se*. What is needed, then, is a market for claims on income flows, so that insurance against adverse shifts in an income flow can be achieved by shorting the market for such flows; investments that diversify into claims on these flows can be made by going long in the market for such flows. Disagreements about the importance of news events will be resolved in the marketplace. So long as individuals in fact buy and sell these policies, then they will see their bank accounts debited or credited in response to this information. The insurance policy will have the appearance of a speculative asset. Creating an insurance mechanism for such income flows is inherently marketizing them, creating a market present value for them.

How, then, are we to create such markets, where the information about income flows is so intangible? Two ways are considered here. One is to create a sort of derivative market, based on an index of observed prices in other (relatively illiquid) markets for claims on the same income stream. Liquid markets in residential real estate might be set up this way; and this example will be pursued below. This method of insuring against adverse information is analogous to that used to provide fire insurance, where the income stream is the flow of rents (or imputed rents) on the property, and the maximum insurance is the present value of this stream inferred from the sale prices of comparable properties. The other way is to create a market for perpetual, or at least long-horizon, claims on an income flow itself, making no use of prices of claims on the flow in any other market. The income flow can be a theoretical construct of statisticians, not the dividend on any tradable asset. Liquid markets in national incomes might be set up in this way, as will be discussed below and in succeeding chapters.

Settlement based on cash-market asset price

Futures, options, swaps, and other contracts that facilitate hedging of risk about the value of claims on flows of future income can be created as derivative markets, where the settlement is in terms of a cash-market price (the price in markets where individual assets are already traded), so long as the cash market price of the claims are observed. The problem that we face in designing such contracts is to make objective observations about such prices. In this section I will discuss contracts that are settled based on appraised values and contracts that are settled based on indices of asset transaction prices. This section will be a brief introduction to the matter, taken up in Chapters 6 through 8, of the theory of price indices for settlement of contracts.

Setting up such contracts is easy if the cash market is already liquid; but our purpose here is to set up contracts in illiquid markets: markets where there is difficult-to-measure heterogeneity of assets traded and infrequent trades of assets of any one description. Futures and options exchanges have already created many contracts that are derivative from liquid markets; to make real progress we must extend from that beginning.

The earliest approach used in futures markets to deal with difficult-to-measure quality of commodities transacted in cash markets is physical delivery at settlement: a futures contract specifies that the shorts who have not closed out their positions as of the settlement date must deliver to the longs a specified amount of the commodity at a specified place. Why is physical delivery, rather than cash settlement based on a price index, advantageous? On some days much of the better quality commodities may change hands in cash markets, on other days less; the average price observed on the cash market may fluctuate from day to day, even when the price of any one grade of the commodity is constant. In this situation, cash settlement may be unacceptable based on an index of cash-market prices. For hedgers, the futures contract could introduce rather than reduce risk. With physical delivery, the quality of items to be delivered can be specified in the contract; experts at the delivery site can study and test the commodity delivered. If only a small quantity is delivered, then expensive testing and evaluation of the quality delivered can be undertaken, something that could never be done cheaply on the entire cash market (see also Garbade and Silber, 1983).

But physical delivery at settlement does not completely solve the problem of costs of evaluation. Because of the costs of evaluation (including the costs of delivery), futures contracts have not been set up to require that shorts deliver a *representative basket* of the commodity, rather that they deliver only one. And they do not require that shorts deliver at a *representative* array of locations, rather that they deliver only at one. Thus, the value of the commodity delivered may not be representative of the broad market for that commodity. If the commodity of the kind and location delivered should turn out to have a different price path than others, then the futures price may turn out to be a poor hedge.

There is an especially good reason to worry that, if only one grade of the commodity and only one location is selected as eligible for delivery, then the price may be particularly unrepresentative of the assets that hedgers hold. This reason is that the process of settlement itself may create a shortage of this kind or grade at this location; there may be a squeeze. So commodities exchanges conventionally allow delivery of any of a number of kinds or grades to be delivered at any of a number of locations, subject to predetermined penalties for delivery of inferior grades or at less preferred locations.

Shorts will always deliver the cheapest kind or grade of the commodity at the location where it is cheapest to deliver. The value of the commodity delivered, then, will be the minimum of the values of the whole array of kinds and locations eligible for delivery. Theoretically, as an index of prices, the minimum price over all kinds is hardly a good one: it is based on the lower tail of the distribution of prices rather than the entire distribution, and the time series of index values may differ from the price path for any given type, since the type priced changes from period to period. The inadequacy of minimum price as an index of all prices has been the identified cause of failure of some futures contracts. Johnston and McConnell (1989) showed that the futures market in Government National Mortgage Association (GNMA) bonds at the Chicago Board of Trade failed after the cheapest-to-deliver bonds became very unrepresentative of the market.

We can do better than use the minimum price when we explicitly construct a price index for cash settlement of futures contracts. The use of physical delivery in futures contracts today is in a sense another of those accidents of history; a holdover from a time when there were not statisticians who could reliably produce a better index than the minimum of a vector of prices. (Physical delivery may still be useful when data to construct an index for cash settlement are costly or unavailable.)

The problem that the value of the commodity delivered may be unrepresentative of the market is exacerbated when the quality of the commodity is hard to define objectively. Then, no penalties can be assessed for delivery of an inferior grade of the commodity, and then the value of the commodity delivered is affected by the availability of deliverable commodities of inferior grades. Because of this difficulty, futures markets tend to exist today in commodities in which the qualities relevant to market price are no more than moderately difficult to measure. Coffee is a good example of those whose quality is hard to measure, but not so hard as to rule out establishing liquid futures markets. The quality of coffee beans is highly variable, and coffee vendors may carry dozens of subtly different varieties. Nonetheless, futures markets have been established based on the subjective tasting (as well as objective measurements) of the coffees at the coffee futures exchanges. Apparently, professional coffee tasters have demonstrated sufficient consistency of judgment that they are trusted by market participants to identify a minimum acceptable grade of coffee for delivery at time of final settlement.

We might imagine copying the institution in the coffee futures market, and providing contracts that were settled with physical delivery of houses, whose minimum quality would be judged by professional appraisers. This would not work extremely well in allowing arbitrageurs to bring cash and futures prices into line at final settlement, because of the great costs of buying and selling houses. These costs of buying and selling are largely a reflection of difficulties in discovering market value. The fees appraisers charge, while substantial, are lower than realtors' commissions partly because the appraisers are not actually committing to buy the property at the appraised price.

One can hardly pass judgment on a house with anything like the ease that a taster can pass judgment on a cup of coffee. The desirability of a given house depends on very many factors, including such things as the current taste in housing styles of the subsection of the population looking in that area, the employment opportunities in the immediate area, the characteristics of the neighborhood, the characteristics of the schools, and even the potential future characteristics of the schools. These factors are so variable that one could probably often find from time to time a *worthless* house to deliver that matches objective specifications such as number of rooms or square feet of floor space. So long as the price of houses delivered bears a consistent relationship with the prices of all houses as a group, this may not be too serious a problem. It would be chancy, however, to base settlement of a futures contract on such delivery, since the relationship between the price of properties delivered and prices of all properties is not one that would seem to be reliable. Even if we had past data on its reliability, these data may not be a solid indicator of future reliability.

As an alternative to physical delivery, cash settlement could be based on an index of appraised prices of properties. To the extent that appraisals are based on prices of comparable properties, rather than just on discounted cash flow or other non-price criteria, the appraisal methods might be adequate to provide an index of prices for settlement. Such appraisal-based indices are, however, expensive to construct, if the appraisals are to be careful. To be representative, the sample they are to be based on must be large. Cost must be part of the reason no such indices are available in the United States for a broad spectrum of properties. The appraisal-based price indices for residential housing in the United States, such as the Coldwell-Banker price index of houses in neighborhoods in which corporate relocations are likely, are based on extremely small numbers of appraisals. The appraisal-based indices for commercial real estate are based on samples, not randomly selected. The best-known appraisal-based index of commercial real estate prices in the United States is the Russell/NCREIF index; it is based on about 1% of US commercial properties.

It will be assumed here that futures contracts will be settled on the basis of indices of prices or rents that rely on market transactions rather than the judgments of appraisers. In the case of real estate, we can (via deeds offices) obtain data on all transactions, and use these to construct real estate price indices that are analogous to the stock price indices. But here we have a problem that, while it afflicts also appraisers, is more severe than it is for appraisers: the same data sources that provide large data sets on transaction prices do not usually provide abundant data on the quality of properties through time.

The most widely quoted index purporting to measure single-family home prices, that of the National Association of Realtors, is just a simple median of house prices sold, making no use of any quality information. Their median shows substantial quarter-to-quarter noise and spurious seasonality. The spurious seasonality occurs apparently because larger or higher-quality houses tend to be sold in the summer. Changes in the median reflect changes in the size and quality of houses sold. There is no published median price of condominiums, and for good reason: a median condominium price would not be a reliable index. The mean number of square feet of condominiums sold in Boston increased from 831 square feet in 1984 to 1,050 square feet in 1991, a 26% increase. A median price would show a spurious increase, reflecting the larger condominiums rather than an increase in price of any given condominium.

To produce a better index of prices of infrequently sold assets, we can use hedonic or repeated-measures indices, or hybrids of the two. These methods rely on the assumed constancy through time of the unmeasured characteristics of assets sold more than once. Those who would try to create indices of coffee prices could not do this, as any given batch of coffee is not sold repeatedly. But, since our interest here is in creating markets for claims on long streams of incomes, we can expect to find repeated sales. These repeated measures methods will be discussed in Chapters 6, 7, and 8 below.

Settlement based on measures of income rather than price

As an alternative to markets that derive from prices in cash asset markets, we can instead set up new markets based on observations of incomes themselves, not on observations of prices of claims on these incomes. These alternative markets do not require the existence of other markets that price present values of cash flows. The alternative markets will create such prices directly.

Two kinds of markets for claims on a series of aggregate income index values are proposed here, perpetual claims and perpetual futures markets. The perpetual claims resemble securities traded at our stock exchanges, and perpetual futures resemble contracts traded at our futures exchanges.

Exchange-guaranteed securities markets for perpetual claims on indices

To investors, i.e. longs, a perpetual claim on an index (such as an index of national income) as defined here is merely a security that pays a dividend proportional to (for simplicity here, let us say equal to) the index value forever. For example, the index value might be national income scaled to $100 in some fixed base year, so that the dividend on the claim would always be proportional to the announced national income. The dividend is paid on every date that a new value of the index is announced. Thus, monthly indices would result in monthly dividends, and quarterly indices would result in quarterly dividends. The investor can buy and sell this security at any time, just as with any other security, and therefore receives both dividend income and capital gains and losses as well.

To create liquid markets for perpetual claims on indices, such as income indices, it is not necessary to find some institution that promises to pay the flow of index values for the duration forever. We need only find an exchange with a clearing house to facilitate a sequence of people transferring the flow of index values to the investors in the perpetual claims. Exchanges can effectively guarantee the continuing flow of index values using the methods of daily resettlement (daily marking to market) and margin requirements already used by our futures exchanges.

With exchange-guaranteed perpetual claims, dividends would be paid by shorts in this market, by those people who freely take the other sides of the perpetual claims contracts. Only shorts pay dividends and so for every long position there is an equal short position. The shorts are not required to hold a claim on an income flow equal to the dividend they must pay, and are under no circumstances required to deliver such a claim. In general, they could not hold or deliver such a claim, since there is generally no other asset that they could hold that can be objectively defined as paying the index value; the purpose of the perpetual claims market is to create such assets.

Shorts, in contrast to longs, are required to put up margin to guarantee their performance. Their margin account is, every day, debited or credited the change in value of the perpetual claim, and when the margin account balance falls below a specified maintenance margin, they are required to put up more margin, or be closed out of their short position through a reversing trade, that is, be forced to buy a long position to cancel their short position. By carefully monitoring margin accounts, the securities exchange and brokerage firms that manage these accounts are able to assure the payment of the index values indefinitely, thereby creating a perpetual security even though no one individual can guarantee paying the dividend forever. The names of the persons who pay the dividends is constantly changing, but the dividend flow is uninterrupted so long as the market continues. Since the net value of all positions (both short and long) in this market is always zero, the market resembles a futures market, but I will not call it this since the value of a contract when it is signed is not zero; to longs it is an investment.

I will stress in this book perpetual claims rather than finite-horizon instruments, for two reasons. First, perpetual claims do not grow shorter-term with time, and so we do not need to have an array of markets in varying degrees of closeness to their maturities. Second, and more important, the perpetual claims represent claims on the entire income stream represented by the index values, so that they price a claim on that cash flow, thereby creating an asset value for the cash flow. Once asset value is discovered, it will be easy for people to derive finite-horizon contracts in terms of the prices discovered in the perpetual claims markets, when desired.[2]

The method of creating markets for perpetual claims described here is modelled after that used for the index participations (IPs) that were traded from May through August 1989 at the American Stock Exchange (AMEX) and the Philadelphia Stock Exchange (PHLX), although the intended application here is very different.[3] The IPs were created to produce exchange-traded stock market basket securities. They were created following recommendations by the Securities and Exchange Commission (1988) following the stock market crash of 1987 that exchange-traded market basket securities be created. Here, the word 'index' in the name of this security refers to the stock price index, not the dividend paid on it. Those who invested in IPs would receive a dividend proportional to the dividends accruing to the portfolio of stocks represented by the stock market index. The dividend was paid regularly by shorts, who put up the usual 150% initial margin to guarantee their payment. The intent of setting up these contracts was to create another market for baskets of stocks; the intent was to compete with the stock index futures markets that already traded baskets of stocks. Since there was no asset underlying the short position, there was some dispute whether the IPs were really securities or futures market contracts, and therefore there was a dispute whether the IPs would be regulated by the Commodity Futures Trading Commission (CFTC) or the Securities and Exchange Commission (SEC). The IPs started trading when they were approved by the SEC, and stopped trading when the US Court of Appeals in Chicago ruled that they were under the jurisdiction of the CFTC rather than the SEC.

I am calling the new securities proposed here perpetual claims rather than index participations for two reasons. First, the term index participation has been usurped by new contracts, since 1989, that replace the shorts with funds holding the assets whose dividend is paid on the index. For the intended applications here, it will generally not be possible to create such funds. Second, the original index participations at the AMEX and the PHLX had a cash-out option, whereby the longs could at any time demand delivery of the stock price index value. When a long exercised this option, there was to be a random assignment to shorts, who would be closed out using the stock price index rather than the IP price.

The cash-out option was made part of the contract out of fear that the IP price might not track the stock price index value well; this cash-out option would prevent the price of the IP from ever falling below the stock price index value. (Shorts were not given a cash-out option, since the exchanges wanted the IPs to be viewed as investment vehicles, and hence did not want longs to be bothered by the potential for cash outs.) In the intended application here, there is generally no observable price analogous to the stock price index on which to base any cash-out option. Rather, the purpose of the perpetual claims markets is to create such a price. The fundamental difference in motivation between the application here and the application at the AMEX and PHLX should be clear: they were not interested in price discovery at all, and clearly hoped that their market would not discover any substantial difference between the IP price and the underlying index value already available elsewhere.

In our stock markets, short sales are relatively unimportant, amounting typically to only a few percent of outstanding value. In contrast, short interest is fundamental to these perpetual claims markets as it is with all existing futures markets. Short interest is more likely to be forthcoming in this market than in stock markets just because there is no other way that people have to hedge their income risks; hedging the risk of income of a stock can be achieved just by selling the stock.

It is to be expected that there will be an enduring demand for (long interest) and supply of (short interest) perpetual claims on aggregate income measures. The long interest in a perpetual claim on aggregate income comes from investors, the short interest from hedgers, who are in turn also investors in different perpetual claim markets. For an example of the use intended here for these markets, consider the hedging use of a market for perpetual claims on national income. Such a market could be used to hedge one's income risk. One merely takes a short position in one's own country's perpetual claim market, so that one pays out the entire component of one's income that is attributable to national income. The hedger would then want to invest the proceeds of that short sale in a long position in some other asset that pays a more secure income, such as a portfolio of perpetual claims on income around the world. For the income perpetual claim market to work well we will have to assume that the hedger will be allowed to make such use of short-sale proceeds. Perpetual futures markets, to which we now turn, are in effect contracts designed to facilitate such use of proceeds.

Perpetual futures

Perpetual futures markets are designed to fulfill basically the same function as the index perpetual claims markets, but they may be more convenient and efficient to deal in for many people. Moreover, they can be designed so that price may be based on a different discount rate than would be involved with perpetual claims markets. One who buys a contract does not pay anything immediately: the contract is not an investment of resources today but a contract that makes one subject to settlement later. The perpetual futures are not derivative from perpetual claims markets; they are not futures on perpetual claims. Perpetual futures markets are alternatives to perpetual claims markets; we do not need to have any perpetual claims markets in order to establish perpetual futures markets. In fact, it is quite possible that only perpetual futures markets would ever be set up. The integrity of the perpetual futures can be guaranteed by exchanges using the same methods of daily resettlement and margin accounts described above for perpetual claims.

With perpetual futures, there would be, every day, a cash settlement, paid from shorts to longs; at time t the settlement s_t is given by:

$$s_t = (f_t - f_{t-1}) + (d_t - r_{t-1} f_{t-1}) \tag{3.1}$$

where f_t and f_{t-1} are perpetual futures prices at days t and $t-1$ respectively, so that $f_t - f_{t-1}$ is the capital gain from $t-1$ to t, d_t is the income index for day t, and r_{t-1} is the return on an alternative asset between day $t-1$ and day t.[4] Here, the letter d, for dividend, is used to represent the income index because it is to be thought of as a dividend on the perpetual claim. The quantity s_t could also be referred to as the daily resettlement or the daily adjustment. Note that the dividend d_t is zero on every day except when the income index is announced, hence the perpetual futures price ought to drop on ex-dividend dates just as do security prices, as, for example, stock prices.

There are two possible interpretations of the cash settlement (3.1) in the perpetual futures market. In the first interpretation, the market combines the daily resettlement of conventional futures with the final cash settlement that in conventional futures markets occurs only on the last day. Recall that in a conventional cash-settled futures market there is, every day t until final settlement, a cash settlement $f_t - f_{t-1}$ in terms of the change in the futures price. Then, on the last day, day T, the final settlement is given not by the change in the futures price but by $p_T - f_{T-1}$ where p_T is the cash-market price index at time T. Thus, for conventional futures markets there are two different settlement formulas, used at different times. In contrast, with perpetual futures we may regard both the daily resettlement and the final cash settlement as occurring every day. By this interpretation of the perpetual futures contract, the term $(f_t - f_{t-1})$ corresponds to the usual daily resettlement, and the second term $(d_t - r_{t-1}f_{t-1})$ corresponds to the final cash settlement (though with perpetual futures it is not final, the contract continues). The term corresponding to final cash settlement looks a little different with perpetual futures: here the income (i.e. dividend) index replaces the cash price index in the settlement formula, and the permanent-income component $r_{t-1}f_{t-1}$ of f_{t-1} replaces the f_{t-1}. By the second interpretation of the settlement formula (3.1), the settlement s_t is just the excess return from $t - 1$ to t between an asset (a perpetual claim) whose price at time $t - 1$ is f_{t-1} (and that pays dividend d_t at time t) and an alternative asset that pays the return r_{t-1} between $t - 1$ and t.

The use of an alternative asset in the settlement formula makes this contract appear rather different from existing futures contracts. But, in fact, existing futures contracts, since they are zero-value contracts when initiated, still must involve an alternative asset in some way. By neglecting the alternative asset in the conventional settlement formula, the exchanges cause the futures price itself to become contaminated by an alternative asset return. The standard textbook description of futures pricing shows that the futures price differs from the cash price by a spread that is affected by the return on an alternative asset, the risk-free interest rate (see the appendix to this chapter). There does exist one futures contract, the Rolling Spot currency futures contract which started trading in 1993 at the Chicago Mercantile Exchange, in which an interest rate (in this case, an interest rate differential between two countries) plays a role in daily resettlement. That contract was designed so as to reduce as much as possible the spread between the spot market and futures market price, so as to make the futures market more of a substitute for the spot market. With perpetual futures, it is even more important that an alternative asset be part of the settlement process; with a perpetual time horizon the effect of neglecting alternative asset returns in settlement would be to drive the futures price dramatically away from the price of the perpetual claim.

It might seem more appropriate to call these contracts perpetual swaps rather than perpetual futures, since the return on a claim on one income flow (the income flow represented by the index) is swapped for the return on another asset. However, the perpetual futures market has a different price discovery function than do swap markets. The price defined in the perpetual futures market is not analogous to those associated with swap markets; in the case where the alternative asset return is the actual return on an asset traded in liquid markets, we expect the perpetual futures price to represent the price of the perpetual claim on the index, a price that may not be observed elsewhere. Moreover, the term swap is conventionally applied to forward contracts with identified counterparties, rather than to contracts traded anonymously at futures exchanges with clearing houses. If one takes part in one swap and does a reversing trade with another swap, there is no clearing mechanism to cancel the positions. With existing swap market institutions, it will be difficult to create perpetual swaps, since the credit risk of the contracting parties cannot be guaranteed forever.

How might the price f_t be determined in the marketplace if there were such a perpetual futures market? The answer to this question depends, of course, on the nature of the alternative asset whose return is used in the settlement formula (3.1). The rate r_t could be a fixed constant independent of time (see the appendix to this chapter). I will generally assume, however, that r_t is a return on a liquid, tradable asset, such as the return on short-term government debt; it might also be the nominal return on indexed debt, if that is available. It is shown in the appendix to this chapter that, under the assumption that there is a liquid market for the alternative asset, no transactions costs, and with a model that allows Euler equation analysis, the choice of *which* alternative liquid asset return to use for r_t does not have any effect on the price f_t; the price f_t is just the market value today of the perpetual claim on the dividend stream d_{t+1}, d_{t+2}, ...

That the futures price will be the same as the market value of the perpetual claim can be seen in a more robust way, without reference to the model that gave rise to the Euler equation, just by assuming the law of one price, that the value of any portfolio equals the value of its components.[5] The value that the market places at time t on the settlement s_{t+1} must be (disregarding transactions costs) zero, since anyone can have that settlement just by entering into a futures contract. It follows that the market value at time t of $f_t(1 + r_t)$ to be received next period equals the market value at time t of $d_{t+1} + f_{t+1}$ to be received next period. Since the market value of $f_t(1 + r_t)$ must equal f_t, f_t must equal the market value at time t of $f_{t+1} + d_{t+1}$. By applying the same argument for f_{t+1} here, proceeding by recursive substitution and assuming that the market value at time t of f_{t+k} goes to zero as k goes to infinity, we find that f_t is the market value at time t of the perpetual claim.

Another way to see that f_t should be the market value of the perpetual claim is to note that by paying f_t one can be assured of receiving the infinite stream of dividends. One does this by buying one futures contract and investing the amount f_t in the alternative asset. In each subsequent period $t + k$, $k > 0$, one invests $f_{t+k} - f_t$ in the alternative asset. By the law of one price, then, and assuming that the present value of f_{t+k} goes to zero as k goes to infinity, the futures price f_t must be the same as the price of the perpetual claim.

Since the choice of the alternative asset, so long as it is liquid, does not matter, one might suppose that it is irrelevant what alternative asset is chosen. But, because of transactions costs, the choice of the alternative asset is not really irrelevant. In considering which alternative asset to use, we must ask which return the market would most desire to swap their specific income risk for. One argument might be that they should most want to swap their return for consumer-price-indexed riskless debt, so that their hedging would eliminate all real risk to them. However, in equilibrium not everyone would be expected to wish to switch to riskless real debt, because it may have a relatively low real return. Since in equilibrium not everyone can get free of all risks, we might expect to see low returns on riskless real debt.

It may be helpful, in gaining acceptance of perpetual futures markets, to use a value of r_t that is always higher than the return on some liquid asset. This would have the effect (see the appendix to this chapter) of shortening the duration of the contract while keeping it perpetual. Contracts that are not quite so long term might be less mysterious to potential traders.

Ultimately, the alternative asset return r_t might best be the return on the world portfolio of aggregate income perpetual futures; this possibility will be considered in the next chapter. But implementation of such a contract must await the creation of a liquid world market in income perpetual futures. It will be assumed in most of what follows that the variable r_t in the settlement formula (3.1) is the return on a conventional liquid asset, such as a risk-free interest rate.

Rational speculative bubbles

Because the net supply both of perpetual claims and of perpetual futures is always zero (for every long there is a short) there appears to be a potential for equilibria where price follows a perpetual growth path unwarranted by fundamentals. To see this, note that adding a term x_t to f_t, where x_t is defined according to $x_t = (1 + r_{t-1})x_{t-1}$, has no effect on the settlement s_t defined in (3.1). So, even if the income stream d_t were constant through time, f_t could still move off to either plus infinity or minus infinity (depending on whether the initial x_t were positive or negative) without affecting settlements. This price path is called a rational bubble: a price path in which there is a deviation from present value that, by its very tendency to grow ever larger, has no effects on excess returns (here, settlements).[6] With perpetual futures, there is no arbitrage possibility that keeps the futures price in line with the cash-market price, because there is no final settlement that ties down the futures price to the cash price at any date. Note that even negative bubbles, where prices fall down perpetually, are possible, so long as the futures exchange allows negative futures prices. The law of one price may not hold, the futures market differing from the cash market just because there is a rational bubble in it.

Since prices appear only in difference form in the settlement formula (3.1), one might say that the bubbles in perpetual futures do not matter; in fact, however, it is conceivable that they might introduce randomness in the price that is not related to the value of the item to be priced. If f_t were contaminated by an additive term x_t where x_t is determined by the difference equation $x_t = (1 + r_{t-1})x_{t-1} + u_t$, where u_t was unforecastable white noise (say, animal spirits, or whims of the market) then neither the expected settlement nor the covariance of the settlement with other returns would be affected, and yet the market would become more volatile. Extraneous volatility might come into the market during each time period through the term u_t, and there appears to be no reason to be sure that such extraneous volatility cannot continue forever to shock the perpetual futures market. Milton Friedman's famous argument (1953) that destabilizing speculation cannot persist indefinitely, since it would mean speculators would lose money on average, does not apply here.[7] We do not have a good theory of these markets that would indicate that such extraneous volatility will not be introduced into perpetual futures markets.

The possibility of rational speculative bubbles may be rather academic, since one might doubt that the market would ever develop an expectation that the price will go off to infinity unwarranted by fundamentals. Unless people feel that the bubble will go on forever, then it will not be rational for it ever to start.[8] Still, to keep the perpetual claim and perpetual futures price informative about the present value of the stream d_t, some kind of price limits may be a good idea. It would certainly seem natural for futures exchanges to prohibit negative perpetual futures prices. Perpetual claims and perpetual futures prices could also be limited to some range around the dividend stream; even if the range is extremely wide, it should rule out rational bubbles. The settlement formula could be rewritten so that the f_t in the formula would be replaced by the limit if f_t in the market went beyond the limit.

Such price limits, of course, do not rule out irrational bubbles, temporary deviations of price due to herd-like or faddish behavior of investors. Irrational bubbles seem altogether a more serious possibility than do the rational bubbles. All speculative investments seem to be vulnerable to such bubbles; perpetual claims or perpetual futures would appear to be no different. Because of the likelihood of such irrational bubbles, we do not yet know how well the perpetual futures price would track the price of a perpetual claim, should both be traded. The example of closed-end mutual funds, that often trade at substantial discounts or premiums from the assets held in their portfolio, and whose dividend stream they share, would suggest that there might be important differences between the perpetual-futures price and the corresponding perpetual-claim price.[9] Even so, the example of closed-end mutual funds also suggests that the perpetual futures and perpetual claims might both work well as hedging media. The discounts and premia of closed-end mutual funds are often fairly stable through time, so that short sales of these mutual funds could have some use as hedging media for the underlying stocks.

Appendix to Chapter 3: Futures markets

An important part of modern financial theory is a class of models based on the assumption that individuals are intertemporal utility maximizers with a present-value utility function. The theory was developed by Merton (1973), Lucas (1978), Breeden (1979) and others. The theory has certain sharp implications about the pricing of perpetual futures markets. Although this theory is of rather limited success in describing actual financial markets (see e.g. Mehra and Prescott, 1985), it is useful to go through its implications, as a way of comparing and contrasting the proposed markets with conventional futures markets.

According to the this theory, households are supposed to maximize expected utility U that is the expected present value of the instantaneous utility $u(\cdot)$ of c_t, consumption at time t:

$$U = E_t \sum_{k=0}^{\infty} \lambda^k u(c_{t+k}). \tag{3.2}$$

where λ is the discount factor, the reciprocal of one plus the subjective rate of time preference.

Let us first, as a review, recall what this theory says about a *conventional* futures market that is one period from cash settlement, so that $s_{t+1} = p_{t+1} - f_t$. The basic Euler equation for a household that considers buying such a contract at time t, that must be satisfied if the household is optimizing, is:

$$E_t(m_{t+1}s_{t+1}) = E_t(m_{t+1}(p_{t+1} - f_t)) = 0 \tag{3.3}$$

where $m_{t+1} = u'(c_{t+1})/\pi_{t+1}$ is the marginal utility of a unit of currency at time $t + 1$ (π_{t+1} is the consumer price index at time $t + 1$). Obviously, if this equation is not satisfied, then the household must be able to improve expected utility by buying more of the contract (if the left-hand side is positive) or selling more of the contract (if the left-hand side is negative). Since households are presumed to be optimizing, then this equation must be satisfied. It follows from the fact that the expected product of two variables is the product of their expectations plus their covariance that the futures price equals the expected cash price plus a risk premium that is determined by the covariance between the marginal utility m_{t+1} of a unit of currency at time $t + 1$ and the cash price p_{t+1}:

$$f_t = E_t p_{t+1} + \frac{cov_t(m_{t+1}, p_{t+1})}{E_t m_{t+1}}. \tag{3.4}$$

While conventional futures prices are determined thus in terms of expected future cash prices, at the same time we can say that, if the cash-market asset is storable, the futures price is determined by today's price and an interest rate. The possibility of zero-cost storage, and the assumption that positive quantities of the commodity are in storage, implies that:

$$m_t p_t = \lambda E_t(m_{t+1} p_{t+1}). \tag{3.5}$$

Moreover, if we have a risk-free asset that pays i_t between t and $t + 1$, we also have another Euler equation:

$$m_t = (1 + i_t)\lambda E_t m_{t+1}. \tag{3.6}$$

From (3.3), it follows that $f_t = E_t(p_{t+1}m_{t+1})/E_t m_{t+1}$ and so, using (3.5), $f_t = m_t p_t(\lambda E_t m_{t+1})$ and hence from (3.6):[10]

$$f_t = p_t(1 + i_t). \tag{3.7}$$

Equation (3.7) is widely interpreted as asserting that futures prices have no relation to expected future prices, being determined only by price today. This can be true as well as futures prices being determined by expected future cash prices because there is a relation, equation (3.5), between the cash price today and the expected future cash price. The relation asserting that futures prices are related to expected future cash prices may be regarded as more fundamental, since it does not rely on the assumption of costless storage opportunities, or even the existence of a cash market at time t. (In our perpetual futures, there is not usually a cash market with any substantial liquidity.)

Let us now inquire how this analysis should be modified for perpetual futures. The basic Euler equation for perpetual futures is:

$$E_t(m_{t+1}(f_{t+1} + d_{t+1} - (1 + r_t)f_t)) = 0 . \tag{3.8}$$

Obviously, if this equation is not satisfied, the household can improve expected utility by buying or selling more of this futures contract; since households are presumed to be optimizing, the equation must be satisfied. This equation is the perpetual futures market analogue to equation (3.3) relating futures price to expected future cash price. It follows that:

$$f_t = \frac{E_t(m_{t+1}(f_{t+1} + d_{t+1}))}{E_t(m_{t+1}(1 + r_t))} . \tag{3.9}$$

Consider the special case where m_{t+1} is uncorrelated with $f_{t+1} + d_{t+1}$ (as might happen if this cash market is small and unrelated to world market conditions) and r_t is a constant (= r). Then the m_{t+1} drops out of (3.9) and we find that the futures price f_t is the present value, discounted by r, of $f_{t+1} + d_{t+1}$. Solving this relation forward, and ignoring the possibility of extraneous bubbles, we would find that the futures price is the present value of expected future dividends discounted by r. Thus, the futures exchange, by setting r, can create markets for present values for any desired discount rate. Setting a high value of r in the contract would mean that the perpetual futures market is relatively short term. The futures exchanges could create an array of futures markets that are forward-looking in various amounts by creating markets with an array of rs. Such an array of contracts might be considered analogous to the array of maturities of futures contracts that are presently offered by futures exchanges, and yet each contract is perpetual. The perpetual contracts do not grow shorter-term with time, so that there is no need for participants to roll over their positions as they expire. However, the conclusion that futures prices will equal expected present values discounted by r depends on the assumption that $f_{t+1} + d_{t+1}$ is uncorrelated with m_{t+1}; more generally, the discount factor in the present value formula would depend on a sort of risk premium that is related to this correlation.

The general assumption in this book will be that r_t is the return on a competing asset that is freely traded. For any such asset, there is an Euler equation $m_t = \lambda E_t(m_{t+1}(1 + r_t))$. Substituting this equation for the denominator in (3.9), we find:

$$f_t = E_t((f_{t+1} + d_{t+1})\lambda m_{t+1}/m_t) \tag{3.10}$$

which is the usual Euler equation for the price f_t of an asset that pays cash flow d_{t+1}, d_{t+2}, ..., that is, of a perpetual claim. It follows that, if there are no extraneous bubbles, the price of the perpetual futures contract would be the same as the price of perpetual claim, were it traded. By this account, it does not matter what alternative asset return r_t is chosen by the exchange for cash settlement, so long as the asset is freely tradable.

4

National Income and Labor Income Markets

For the purpose of hedging risks to standards of living, the logical place to look first would be to markets for claims on total income. It is curious that such markets do not presently exist, and that they have apparently never even been proposed before. What economic variable could matter more to people, what economic risks could count for more, than their total incomes?

By making it possible to hedge the capital value of a stream of aggregate income, perpetual claims or perpetual futures markets, long-term swap markets, or retail analogues of these would facilitate management of the kind of longer-run income risk that really matters to individuals and organizations. Nations or other groupings of people could use such markets to insure themselves against the prospect of a declining standard of living, against the prospect of relative poverty. By hedging such risks, these macro markets would allow the natural tendency for convergence of incomes to reduce inequality of incomes, by removing the shocks that disperse incomes. Thus, the establishment of such markets might make significant progress, in the long run, toward equalizing wealth across nations, regions, and categories of people and, consequently, across individuals themselves.

There could be markets for hedging the risk of fluctuations in aggregate income, national income or aggregate labor income, for each country of the world. Moreover, since national boundaries are not always the best criteria for defining income aggregates for hedging purposes, markets may also be established for other regions. Markets for income could be divided up in yet different ways. We could have aggregate income markets for flows associated with human labor characteristics, with occupations, and with investments in human capital.

Most of people's income is labor income, a return for their services, and so creating markets for claims on total income means for the most part creating markets for claims on labor income. A market for claims on total income would enable them to hedge their labor income risk along with other income risk. A market for claims on labor income might, on the other hand, be used along with markets for claims on other components of total income to hedge total income as well. Whether we want to set up markets in aggregate income or in labor income, or both, is not the major issue that concerns us here. Presumably, that choice would depend on the existence and liquidity of markets for other components of income, and on issues of convenience and simplicity for hedgers.

Market structure and associated institutions

Let us consider explicitly some possible hedging arrangements in perpetual claims or perpetual futures markets for national incomes. Let us begin by imagining that the household deals directly in the perpetual markets itself, even though most such dealings would probably be intermediated by retail institutions. Suppose, for illustration, that a household's income correlates perfectly with a macroeconomic aggregate represented by a perpetual claims market. A hedging household would then sell an amount of perpetual claims so that the amount it pays in dividends exactly equals its income. It would then use the proceeds of the short sale to invest in an alternative asset with a more stable dividend flow, such as a portfolio of perpetual claims on incomes around the world. If the household subsequently leaves these positions undisturbed, then it has in effect swapped its income flow for another more stable flow of income.

When the household first takes these positions, the net value of its positions is zero. The household will see the value of its position rise and fall with the market in perpetual claims on its own income and the market for the alternative asset. Should the value of its position become too low, then the household will be subject to a margin call, and must transfer assets to a margin account. Such transfers are necessary to guarantee performance, since an individual whose combined positions acquires a negative value would have an incentive to walk away from them. An individual whose combined positions acquire a positive value has in these combined positions an asset that can be sold at any time, but presumably there is no need to sell, if the combined positions continue to provide a stable income flow. The asset would probably usually be bequeathed to subsequent generations.

The same hedging could be done with perpetual futures, since perpetual futures, under the assumption that perpetual futures prices are the same as perpetual claims prices, consist essentially just of a long position in a perpetual claim and a short position of equal value in the alternative asset. One begins just by taking a short position in the perpetual futures market. As time goes on, one then continues to take the short position in the perpetual futures market and reinvests the combined capital gains into the alternative asset, and continues to consume the dividend on the alternative asset indefinitely.

Those who live in countries whose national incomes are very uncertain may or may not have to pay, in effect, an insurance premium in the form of an expected loss on their income-hedging positions. This expected loss is the analogue of the losses due to backwardation in conventional futures markets. If the price of a claim on a country's income is both uncertain and substantially correlated with the aggregate world market, then there would be reason to think that the expected return on a perpetual claim on its income would be larger than the average expected return, its risk premium would be larger than average, and so the combined positions of the hedger would tend to lose money on average. On the other hand, if the price of a claim on a country's income is relatively certain and not very correlated with the aggregate world income, then there would be reason to think that the expected return on a perpetual claim on its income would be smaller than average; there may be hardly any risk premium implicit in the price of the perpetual claim. Individuals in such a country may participate in the hedging markets (shorting perpetual claims on their own incomes and investing in a portfolio of perpetual claims around the world) more for the expected return that they get for sharing in the world's risks than for the reduction of their own risks.[1]

It should also be noted that, if indeed everyone who is short in a regional market desires to be long in the world market, then it might be advantageous to create a perpetual futures contract with the world macro futures market return as the alternative asset return in the settlement formula. Each futures contract would then be, in effect, a swap of the return on a specific claim on income for the return on a claim on world income. If the return r_{t-1} on the claim on world income were defined in such a way that the weights corresponding to the different specific incomes corresponded to the number of shorts in the macro futures contracts in those specific incomes, then the sum of the settlements paid to shorts would be identically zero, and the market might exist with only short sales. (Effectively, a long position in one market would be equivalent to a short position in all other markets.) Using n_{it-1} to denote the total contracts in market i at time $t - 1$ and N as the number of futures markets, we would define r_{t-1} as:

$$r_{t-1} = \frac{\sum_{i=1}^{N} n_{it-1}(f_{it} - f_{it-1} + d_{it})}{\sum_{i=1}^{N} n_{it-1} f_{it-1}} . \tag{4.1}$$

A settlement formula based on such an alternative asset return must, however, await a world market for macro futures. Probably, the practical way to start macro futures is to settle on the basis of a world interest rate rather than something so ambitious as defined in equation (4.1). Moreover, if such markets were the only liquid markets for national incomes then there would be no way to invest liquid wealth in perpetual claims on world income; we would still need other markets that make this possible.

In practice, of course, a household's income will not correlate perfectly with national, regional, or any other income aggregate. The household may then wish to hedge less than fully on any one such market, and may instead want to hedge partially on a number of income markets, each of which has some correlation with its own income flows.

Problems enforcing payments of losses

An important problem faced in markets for claims on aggregate incomes is that the hedging household that incurs losses may wind up unable to meet margin calls. This would happen if the household's own income rose far enough to wipe out its liquid assets. If the household were to deplete its store of liquid assets, then it might find that its income would no longer be hedgable, because it finds it difficult to sell claims on its future income to come up with margin. This possibility does not completely vitiate the hedging function of the macro markets; it would only mean that not everyone can use the macro markets at all times to hedge.

The problem that losses in perpetual markets for income may exceed liquid wealth could be reduced if it were possible for individuals to sign contracts to pay from their own future income in exchange for cash today to meet margin requirements. In practice, of course, the ability of individuals to sell claims on their future income is quite limited, partly because of the difficulty of enforcing the contract and partly because of personal bankruptcy laws. Friedman (1962), who advocated allowing, for educational purposes, private sales of such shares in future income, thought that there would probably be irrational public condemnation of such sales.

Setting up institutions to allow people to sell claims on their own future incomes is, of course, analogous to creating a market for such claims, but it is not the same as creating a liquid market for a perpetual stream of future income. The household would need to do no more than find someone who knew it well (e.g. a local banker) who would be willing to buy a claim on its future income. People could sell such claims in order to produce cash to meet margin requirements in hedging markets (or do the equivalent of such hedging through intermediaries). Such claims would be inherently heterogeneous, differing in payout structures and risks of default, and so an index of prices in that market may not work well to cash settle futures contracts.

Our laws and institutions should be changed to make it easier for people to sell claims on their future incomes to come up with margin for hedging purposes. This does not mean that we should eliminate laws and institutions that help protect people, by preventing them from frivolously consuming their future income from their own foolishness, thereby enslaving themselves, only that the laws should allow them to sell claims for the sensible purpose of hedging. Governments are ultimately able to enforce payments by individuals (witness our income tax laws), and they should be able to do that with payments occasioned by losses in macro markets.

People will be unable to trade in claims on future incomes unless it is believed that governments will enforce payments in the future. When people in one country have made heavy losses in a macro market to another country, will they continue to support a government that makes them pay? Whether they will is hard to say for sure. There is some reason for optimism. Most countries pay their international debts today, even when they are in deteriorating economic circumstances. The debts incurred as a result of losses in income markets would be incurred in relatively improved economic circumstances, so the macro market losses would not be hard to bear. Moreover, social psychologists and political scientists have described a strong sense among most people of the importance of proper procedure among their leaders. Although people will lobby for their special interests, they will usually accept major losses if they feel that the enforcement of the losses is proper, and if their leaders are essentially fair (see e.g. Leventhal *et al.*, 1980; Tyler and Caine, 1981; Folger and Martin, 1986).

In practice, changes in laws and institutions to facilitate people's selling claims on their own future labor income are not likely to occur for some time.[2] Still, some use can be made of aggregate income markets to hedge income risks. If a household finds it difficult, because of inability to commit to pay in the future, to hedge all of its income risk, it can, so long as it is able to commit to pay from some of its future income flows, such as its property income, still hedge some of it. The household could follow the strategy of hedging a fraction of its income in the macro market, and raising the fraction if its income should go down, lowering it if it goes up. If income should go down, the household has winnings in the macro market that would increase its ability to meet margin calls, and so the household is better off and so does not need to hedge so much. Such a strategy may be called a dynamic portfolio strategy for replicating an out-of-the-money put on income. The household is, if it responds in the proper manner in its hedging to its income fluctuations, effectively buying a put on margin that creates a floor on the present value of its income below the present value of its current income. The effective price of such a put would be fairly small if the floor is sufficiently low.

Thus, even though the household may be unable to stabilize its income completely because of inability to commit to pay in the future, it can buy insurance against disastrous decreases in its income.[3]

Other users of aggregate income markets

The hedging ability provided by aggregate income markets may also work towards reducing risks to standards of living through other, less direct channels than the direct hedging of individual incomes. Among the participants in the macro markets would logically be firms hedging their costs of producing. The aggregate income measures in a particular country are correlated with the costs firms must incur in hiring people and other resources there. By hedging the risk of changes in these costs or dealing with intermediaries that insure them against such changes, they can exploit international cost differences more effectively. The macro markets would therefore facilitate international capital flows in response to production cost differentials, encouraging industry to move to places where it is most productive. It is well documented that the barriers to such capital flows are quite significant. Capital investments in each individual country are today determined largely by the total saving in that country, reflecting serious inefficiencies in the international allocation of capital.[4]

Markets in actual or full-employment income?

Should an income risk-management contract be written in terms of total actual income of a grouping of people, or their full-employment income? Changes in full employment income represent changes in rates of earning income. When considering labor income, then, we may regard full employment income as proportional to the wage rate. Actual income and full-employment income are conceptually different because of time variation in their employment or utilization rates. There may even be room for markets in both.

It might seem perhaps most natural to create a market for actual income. Consider labor income. One might well argue that workers would want to be insured against income fluctuations due to unemployment as well as to other income fluctuations. Insurance companies could retail private income insurance policies that insure against the present value of a lifetime of frequent low-employment states—although some means of handling the moral-hazard problem would be necessary. Assuming that insurance companies do manage to market private unemployment insurance policies, they will then want to be able to hedge the aggregate risk that they will have to make up the lost income of these people; for them, a contract in aggregate labor income may be ideal.

However, the fundamental problem is that declines in employment status and numbers of hours worked are also offset by another benefit—declines in effort expended and increases in leisure available. For this reason, workers are not likely to view declines in income due to declines in hours worked as the same as declines in the wage rate. It is conceivable that for them there is no cost at all to declines in hours worked so long as the wage rate remains constant. Indeed, changes in the number of hours worked may largely represent changes in the desire for work. This is especially likely for long-term changes, and the perpetual markets are sensitive to long-term changes. We would not want to see workers receiving a settlement on their perpetual contracts just because they are working less because of a decision to spend more time in leisure. The firms on the other side of the contracts, hedging labor costs, would in effect be forced to pay workers the same amount even when they chose to work less.

To some extent, secular trends in hours worked per week may reflect some things other than changes in tastes. There may be a secular down-trend in hours worked per week as the standard of living rises; this down-trend may have a predictable relation to the level of income, and hence, because there is virtually no uncertainty about this relation, have no effect on the functioning of hedging markets.

Defining contracts in terms of wage rates instead of labor income may have advantages for firms using the hedging markets to manage their labor cost risks. We may indeed want to tailor a market for such hedging use; we could, for example, create long-term hedging markets, such as perpetual claims markets, in the employment cost indices published by the US Bureau of Labor Statistics.

Measurement issues

Many existing aggregate income, earnings, or employment cost indices may not be ideal for the purpose of cash settling macro market contracts. These measures may not accurately represent the path through time of real individual endowments that people want to hedge.

National incomes are affected by births and deaths, immigration and emigration; to the extent that variations in these may reduce the effectiveness of national income in representing incomes of individual people, and to the extent that these variations are unpredictable, they may compromise the hedging function of the hedging market. Basing trade on per capita national income, rather than national income *per se*, may alleviate these problems somewhat, but not completely. Changes in the age distribution of the population may affect per capita national income without anyone's income being affected. The kinds of people who emigrate and immigrate may have different levels of income.

Consider, moreover, the possibility of setting up perpetual claims markets in the median earnings per week for occupational categories as published by the US Bureau of Labor Statistics. The earnings may vary through time because of the composition of the labor force in the defined category. High-wage industries within a certain category may grow or decline through time, thereby causing changes in the median wage paid even when no individual suffers a change in pay. Moreover, the occupational categories were not chosen to represent individual endowments, so that people can freely shift from one to another. For example, the median weekly earnings within the category of medical workers reflects, since it is the median, the earnings of the relatively low-paid individuals in that field, not the medical doctors. Most of these people have not made the heavy investment in training that doctors have, and many of these people will switch to another occupational category should wages for medical workers fall.

It is essential, then, to construct income, earnings, or wage indices that follow through time the earnings of specific categories of people or investments, using methods like those described in Chapters 6–8 below.[5] Moreover, these categories should be representations of endowments, categories that are not easily changed.[6]

The most obvious category is the nation; citizenship is difficult to change, and the extent of immigration and emigration is limited. It is only for this reason that national income should have any priority over other income aggregates in the establishment of macro markets. There is nothing political about impersonal financial markets, and so political boundaries may have limited meaning for them. We should also consider other income aggregates.

Markets should be opened for incomes in specific occupational categories that are difficult or expensive to change, such as medicine or law. There is already a sort of forward market in medical careers, whereby doctors try to hedge the risk of adverse changes in their incomes.[7] The total value generated in existing cash markets for medical incomes is not small, certainly comparable to the income generating in cash markets where there are active futures markets.

A problem with establishing occupational income markets is that there are vastly many sorts of occupations. Presumably, there should be some efforts to group together occupations whose market wages move together. Reich (1992) has claimed that jobs can be grouped into three broad categories: routine production services, in-person services, and symbolic-analytic services, and argued that in today's international economy a person's category may matter more than the person's country in determining income prospects. Much more research could be done in defining such categories.

The methods currently used to produce employment cost indices might be improved if these indices are to be used to settle risk management contracts. Greater attention could be paid to controlling for the change in mix of labor hired, a repeated-measures framework that follows individual workers might be appropriate.

Measuring uncertainty about present values of incomes

How much short-run variability would there be in national income present values, variability that would cause price volatility in perpetual claims or perpetual futures markets? If a hedging market is to be successful, there must be enough action in price to interest traders and enough uncertainty to concern hedgers.

National income and other income measures often behave rather smoothly through time. This smoothness might suggest that there is little uncertainty about the present value of aggregate incomes. If the income series is very smooth, then future values can be forecast very well by pure extrapolation. If, then, there is in fact little uncertainty about future values, then there is little incentive to hedge, and, moreover, the price of the asset will not be very volatile.

The smoothness of an income series is no proof that a perpetual claim on incomes will be smooth. Dividends paid on corporate stocks also behave fairly smoothly through time (companies try to smooth nominal dividends), and this does not result in smooth prices of stocks.

Of course, if incomes are *really* smooth, so much so that an extrapolation of incomes is very successful in predicting future incomes as far into the future as is important in the context of the present-value formula, then there will not be much uncertainty about the present value of future incomes, and so not much volatility in a market for perpetual claims on these incomes. The same would be true if there were some other information variables, apart from incomes themselves, that are quite successful in forecasting incomes as far as is relevant for the present-value formula. In these cases, and assuming that price is the present value of optimally forecasted future incomes, there will not be much variability of returns. In the extreme case where the future is known perfectly, then there will be no uncertainty at all about future incomes, and hence no variability in returns on claims on incomes. The question that one must address in viewing the income series is to what extent the present value of this series is forecastable.

Results with individual countries

To investigate the forecastability of present values, and to see the connection between this forecastability and the variability of prices in perpetual claims markets, I will use a log-linearized form of the present-value model that John Campbell and I developed (1988; 1989).[8] The model is described in the appendix at the end of this chapter. With this model, we can produce estimates of the standard deviations of returns of perpetual claims on incomes, quite a different animal than standard deviations of changes in national incomes. Of course, our estimated standard deviations are only as good as our model. As discussed in the appendix, the method of estimating standard deviations of returns with this model seems to produce substantially lower predicted standard deviations than we actually observe in existing stock markets, and so the estimated standard deviations appear to be conservative.

The method of estimating the standard deviation of returns was applied to each of the countries of the world for which the Penn World Table Mark 5.5 (see Summers and Heston, 1988; 1991) had real gross domestic product data for the years 1950–90.[9] In applying this method, the value of the annual discount factor ρ was taken to be the same as that chosen by Campbell and Shiller (1988; 1989) for the stock market: 0.936.[10] Alternative estimations with the number of lags in the autoregression k equal to 5, 10, and 15 years were undertaken; since the results were broadly similar only the results with $k = 10$ are displayed here.

Table 4.1, column 1, shows the estimated standard deviations of these theoretical returns for 54 countries in hypothetical real gross domestic product markets, computed from equation (4.9) in the appendix to this chapter. We see from the table that the standard deviations of returns in the national income perpetual claims market is sometimes quite large, comparable to that of really speculative assets.

The estimated variances of returns might, of course, be lowered if other forecasting variables were used, as for example, the forecasting variables used by Barro (1991). Including a simple time trend in the forecasting regression for dividends might also lower variances dramatically. But, as argued above, we can never be sure how much the forecastability of national incomes can really be improved, since there is ultimately a degrees-of-freedom problem; there are too many potential forecasting variables.

The standard deviation of return in this market for claims on gross domestic product is extremely low for the United States, the United Kingdom, and some other English-speaking countries. It is also low for some, though not all, European countries. This means that, over this sample period, the present value of national income has been quite forecastable. That the United States and the United Kingdom have such low standard deviations of returns seems unfortunate, given that these countries are likely testing grounds for many innovative risk-management contracts. On the other hand,

Table 4.1: *Statistics on Theoretical Annual Returns and Values: Perpetual Claims Markets in Per Capita Real Gross Domestic Products*

Country	(1) Standard deviation[a] (%)	(2) Value of country[b]	(3) R squared[c]	(4) Beta[d]
Argentina	9.86	2,460	0.05	1.14
Australia	3.18	4,479	0.30	0.92
Austria	3.18	1,608	0.44	1.09
Belgium	3.71	2,201	0.47	1.34
Bolivia	5.45	178	0.14	1.08
Brazil	5.86	9,776	0.12	1.09
Canada	2.56	8,155	0.31	0.73
Chile	4.90	978	0.29	1.38
Colombia	3.42	1,783	0.30	0.99
Costa Rica	6.35	170	0.19	1.46
Cyprus	3.22	98	0.31	0.94
Denmark	3.56	1,201	0.14	0.71
Dominica	6.75	285	0.00	0.20
Ecuador	5.89	450	0.00	0.07
Egypt	3.33	1,669	0.01	–0.13
Finland	3.43	1,288	0.06	0.44
France	5.27	13,570	0.44	1.83
Germany (West)	4.39	15,638	0.43	1.44
Greece	7.87	1,107	0.12	1.46
Guatemala	6.13	316	0.27	1.67
Guyana	11.04	14	0.03	0.98
Honduras	4.58	120	0.23	1.16
India	5.10	14,619	0.01	–0.20
Ireland	2.79	507	0.15	0.57
Iceland	4.53	59	0.02	-0.31
Italy	5.08	13,543	0.43	1.71
Japan	8.38	29,934	0.36	2.67
Kenya	4.40	355	0.00	–0.02
Luxembourg	2.40	111	0.10	0.41
Mauritius	6.20	110	0.01	–0.28
Mexico	6.01	7,608	0.12	1.08
Morocco	3.01	855	0.04	0.30
Netherlands	4.72	3,193	0.34	1.45
New Zealand	2.85	684	0.32	0.85
Nigeria	10.74	1,179	0.08	1.54

Country	(1) Standard deviation[a] (%)	(2) Value of country[b]	(3) R squared[c]	(4) Beta[d]
Norway	2.21	1,049	0.01	0.09
Pakistan	3.07	2,612	0.01	0.12
Panama	7.08	97	0.00	0.25
Paraguay	6.11	157	0.00	0.17
Peru	11.06	984	0.00	−0.29
Philippines	3.68	1,602	0.01	0.15
Portugal	7.00	1,115	0.24	1.81
El Salvador	9.20	209	0.03	0.89
South Africa	8.68	2,272	0.08	1.29
Spain	6.60	5,631	0.09	1.04
Sweden	3.75	2,194	0.00	0.12
Switzerland	4.33	1,986	0.35	1.33
Thailand	5.02	3,125	0.22	1.21
Trinidad and Tobago	8.75	157	0.01	0.49
Turkey	3.59	3,151	0.01	−0.16
United Kingdom	1.14	13,616	0.05	0.13
United States	1.62	81,044	0.49	0.59
Uruguay	4.44	236	0.11	0.76
Venezuela	9.18	2,385	0.01	0.35

[a]Of return in market for perpetual claim on GDP.
[b]1990 expected present value of real GDP, 1990 $US billions.
[c]Country return regressed on world market return.
[d]For return in perpetual claims market for GDP.

Source: Author's calculations as described in text using annual per capita real gross domestic product and population data 1950–90, both from the Penn World Table Mark 5.5. For data description, see Summers and Heston (1988; 1991).

even with a standard deviation of only 1.62%, the total dollar value of the hedgable risk in the United States far exceeds that implied by this method for the stock markets in the United States, since the present value of national income is over an order of magnitude larger than the present value of corporate earnings. We may also note that the fact that the period 1950–90 was a highly forecastable one does not mean that in other time periods gross domestic product is as forecastable. Indeed, when the sample period is extended for the United States, where the autoregressive model is estimated using per capita real GNP data from 1889 to 1992, then the standard deviation of returns (estimated with the identical procedure with ten lags in the autoregression but using the longer sample period) from 1900 to 1992 is 4.72%.[11] We might also infer from the experience of other countries that the United States is potentially vulnerable to large shocks as well. Indeed, the recent public concern with declining competitiveness of the United States in the world markets suggests that a perpetual claims market in the US national income might indeed show substantial volatility.

World market return and betas

A comparison of the variance of the world portfolio with the variance of individual country market national income perpetual claims market returns will allow us to judge the extent of unhedged risk of national incomes. If the standard deviations of returns on the world portfolio are smaller than those of returns on country portfolios, then, assuming that there are now no markets to handle such risk, we show an advantage for countries using perpetual claims or perpetual futures market to swap, in effect, their individual returns for the world returns.

To make this comparison here, the first step was to compute the expected present value of the gross domestic product for each country, to serve as weights in the computation of return on the world portfolio. The expected present value was produced for each country by dividing the real gross domestic product for that country and that year by the theoretical real dividend-price ratio for that year (the ratio computed from equation (4.7) in the appendix to this chapter, an equation that uses as inputs lagged growth rates of real income for that country). The 1990 value of this expected present value in 1990 dollars is shown for each country in column 2 of Table 4.1.[12] Note that the expected present values, the values of the countries, are much higher than the usually defined national wealth for the countries. For example, the value of the United States is shown there to be $81 trillion in 1990, much higher than the $18 trillion domestic wealth in the Flow of Funds Accounts (Board of Governors of the Federal Reserve System, 1992).[13] This discrepancy is, of course, what we would expect, given that, historically, roughly three quarters of national income is labor income, whose capitalized value is not included in conventional measures of wealth.

Having generated real returns time series and the value weights for each of the 54 countries, we compute a world real return series by forming a weighted average of these returns; the weight given to each country in a given year proportional to the value of the outstanding perpetual claims market for the gross domestic product of that country. Of course, this estimated world return is not likely to be a very accurate indicator of actual returns in markets for world income, because of our lack of knowledge of actual information sets and because of our inability to model speculative price changes. Still it is possible to compute a rough indication of world returns that will give us a crude sense of how variable such returns might be, and allow us to get some idea of how much returns in different countries might correlate with each other.

Fig. 4.1 shows a plot for the years 1961–90 of this world real return series. The standard deviation of the real return (after a degrees-of-freedom correction that was employed because this real return is based on the residual of a regression) is only 1.90%, suggesting that much of the risk of national incomes can be diversified away internationally. This suggests that the world market risk premium will be very small, and that there might be little backwardation in perpetual futures markets. Still, we should not understate the world market volatility; there have been some fairly important market turns, notably after the 1973 oil crisis, when there was a two-year return of –7.36%.

Although world market return is probably not well measured by this method, it is still instructive to compute country betas, to get some idea of how much countries might differ in their exposure to world risk. Betas were computed here for the perpetual claims returns of each of the 54 countries, by regressing the country return on the market return. The R squared of each regression is shown in Table 4.1, column 3; note that the R squared is usually quite low, reflecting the large idiosyncratic risks that countries are now bearing. The estimated betas appear in Table 4.1, column 4. Note that there is much variability of the betas across countries. The beta for the United States is only 0.59. One might have expected higher, since the United States has a major weight in the world real return series; it accounted for 30.7% of world

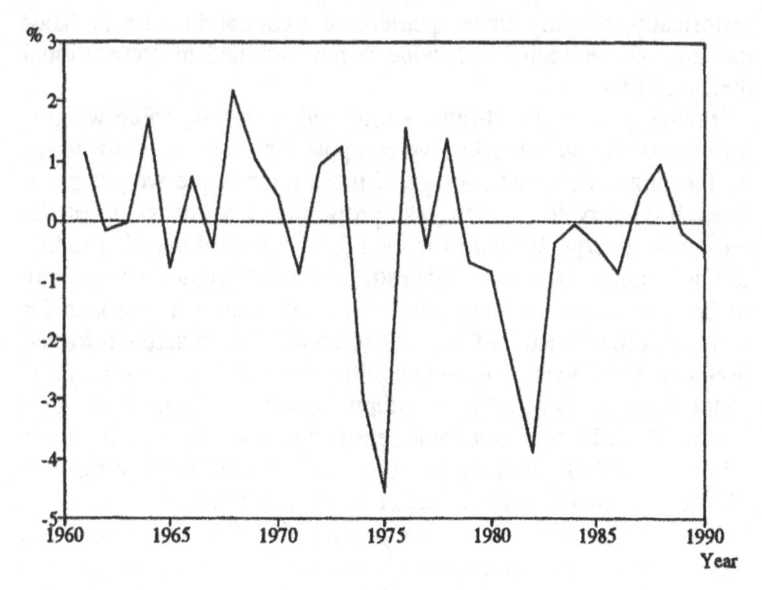

FIG. 4.1 Estimated World Income Market Returns, Demeaned, 1961–1990.
Source: Calculations described in text from Penn World Table 5.5 data on real gross domestic products 1950–1990.

market value of $264 trillion in 1990 as computed here (omitting, of course, the Soviet Union and China and some other countries). But the standard deviation of the United States return is estimated here to be quite low, and so it has only a small direct effect on the world return.

Cross hedging

Could market participants take a position in some other financial market, other than macro markets, to lay off their aggregate income risks? It is conceivable that the price in a perpetual claims market would be very closely correlated with the price of, say, corporate stocks. Then there would be no need to establish the new markets.

It is not a trivial matter to resolve to what extent the cross hedging might work, since we do not now observe the market price of a claim on a stream of aggregate income. The best we can

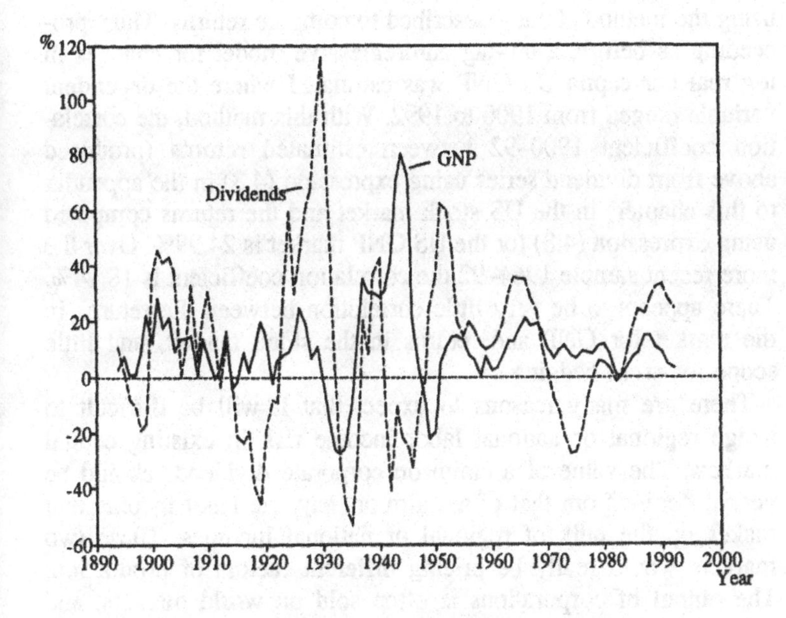

FIG. 4.2 Growth rates for five years ending in year shown for real dividends (dashed line) and real per capita US GNP, 1894–1992.
Source: Standard and Poor's, Shiller (1989) and US Dept. of Commerce.

do is to make inferences based on a comparison of the stream of aggregate incomes with the stream of dividends on existing financial assets, or to use the theoretical prices of the claims described above. We must be careful to compare flows with flows or stocks with stocks, and not compare flows with stocks.[14]

Examination of existing long time series of US real per capita gross national product with real dividends accruing to the Standard and Poor Composite Stock Price Index reveals that the two series have virtually no relation. Five-year growth rates for both series are plotted in Fig. 4.2 for each year 1894–1992. The correlation 1894–1992 of five-year growth rates in real dividends accruing to the Standard and Poor Composite Index with five-year growth rates in real per capita GNP is only a minuscule 2.81%; the correlation of the five-year growth rates in real earnings with five-year growth rates of real per-capita GNP is only 20.66%.[15]

We can also compare estimated returns in the market for US GNP with estimated returns in the market for corporate dividends, using the method already described to compute returns. Thus, proceeding as before, a ten-lag autoregressive model for changes in log real per capita US GNP was estimated where the dependent variable ranged from 1900 to 1992. With this method, the correlation coefficient 1900–92 between estimated returns (produced above from dividend series using expression (4.8) in the appendix to this chapter) in the US stock market and the returns computed using expression (4.8) for the US GNP market is 24.99%. Over the more recent sample 1964–92 the correlation coefficient is 18.14%. There appears to be very little correlation between the returns in the market for GNP and returns in the stock market, and little scope for cross hedging.

There are many reasons to expect that it will be difficult to hedge regional or national labor income risk in existing capital markets. The value of a claim on corporate dividends should be very different from that of a claim on, say, the labor income that makes up the bulk of regional or national incomes. These two markets would really be pricing different factors of production. The output of corporations is often sold on world markets, and reflects international conditions. Corporations are increasingly international and move their operations around the world. Labor is relatively immobile, much of it engaged in activities that are not directly connected with corporate activities.

For most of the countries included in the econometric study above, stock markets are less important to their respective economies than is the case in the United States, and in many cases stock markets are nonexistent. For these countries, there is even less likelihood that cross hedging on any existing markets could be effective in reducing income risk.

Interpretation of results

The evidence presented here that there is much income risk to be hedged in claims on gross domestic products suggests that national income or labor income hedging markets will be very valuable. For some of the countries that we studied, the variability of returns is estimated to be about the same as that calculated for the US stock market (recall that this method seems to imply that actual stock market volatility should be lower than it is); for many of them the risk is at least comparable to the stock market risk. We have seen that existing capital markets do not seem to allow cross hedging of this risk. The macro markets that are proposed in this chapter are logical directions to look for reducing the effects of such variability on people's lives.

It may seem surprising that the volatility of prices of claims on national incomes is so often comparable to the volatility of stock prices. Stockholders are residual claimants, receiving a claim on the profits of the corporation after everyone else has been paid off. It would seem that such claims would be much more volatile than claims on aggregate income flows. In fact, however, although it would seem that stocks should be much more volatile, historically they have not been. Siegel (1994) has shown, using US data from 1802 to 1992, that 20-year real returns on corporate stocks have actually been less volatile, in real terms, than corresponding returns of either long-term bonds or treasury bills. It is actually the residual claimant who appears to shoulder less risk.

The evidence presented here complements earlier evidence that risk sharing internationally is not optimal. Such evidence has been found by comparing consumption growth patterns across countries, on the premise that if there were complete sharing of risk internationally, there would be perfect correlation across countries in consumption growth rates. Backus *et al.* (1992) found that correlations since 1960 of contemporaneous consumption (filtered to eliminate low frequency movements, below business cycle frequency) between the USA and eleven other major nations were quite low. Only Canada had a correlation greater than a half; three were negative.[16] However, the low correlation might be attributable to country-specific taste shocks affecting saving behavior rather than to inability to hedge (Stockman and Tesar, 1990). That such taste shocks are plausible may compromise the ability of consumption correlations to reveal the extent of unhedged risk.

Atkeson and Bayoumi (1991) attempted to find some evidence whether labor income fluctuations are in fact hedged in existing capital markets. They used time series regressions for each of various regions. In each regression, changes in per capita income from capital in that region was the dependent variable. The independent variables were changes in per capita income from capital for a broader aggregate of regions, changes in per capita income from labor in that region, and changes in capital product per capita for that region. The last independent variable is measured from the production side of accounts rather than from the income side. The coefficient of the change in per capita income from labor term is of interest here: if people in that region were using the capital markets perfectly to hedge, then we might expect a coefficient of the change in per capita income from labor in that region of minus one. Atkeson and Bayoumi ran these regressions with the constraint that the coefficients were the same across regions. When the regions were the states of the United States of America 1966–86, the estimated coefficient of the change in per capita income from labor was –0.004. Although this coefficient was significant at conventional levels, it was far from minus one; it was really inconsequential in magnitude. The coefficient of change in aggregate per capita income from capital was 0.983, virtually 1.000, the coefficient of the change in capital product per capita 0.022, virtually zero. When the regions were six members of the European Common Market (Germany, France, the United Kingdom, Belgium, the Netherlands and Greece) 1970–87, and the aggregate across regions was the sum of the incomes for the six countries, the coefficient of the change in per capita income from labor was –0.045. This is a little more substantially negative than was the case for the individual states in the United States, but still very far from minus one. Their evidence indicates a near-total failure to hedge income risk in existing markets, further confirming the potential value for macro markets in national incomes.

Appendix to Chapter 4: Econometric methods

The log-linearized present value model, or 'dividend-ratio model' (Campbell and Shiller, 1988; 1989) is attractive because it deals in percentage changes rather than levels of variables entering the present-value relation; the former are more plausibly assumed to be stationary stochastic processes. In place of price, which may not be stationary, the model is cast in terms of a ratio, the dividend price ratio. We define δ_t as the (demeaned) natural log of the dividend–price ratio, equal to $d_{t-1} - p_t$ (demeaned) where d_{t-1} is lagged log dividend and p_t is log price. The dividend is lagged one period since both δ_t and p_t are supposed to be measured and publicly observed at the beginning of time t. The dividend d_t paid over all of time t is not available at the beginning of time t to form the numerator of the ratio; indeed, conventionally defined dividend–price ratios use lagged dividends for the numerator. The basic approximation can be represented as that δ_t is the expectation of the present value δ_t^* of the future changes in log dividends:

$$\delta_t = E_t \delta_t^* \tag{4.2}$$

$$\delta_t^* \equiv -\sum_{j=0}^{\infty} \rho^j \Delta d_{t+j}\,, \quad 0 < \rho < 1 \tag{4.3}$$

where Δd_{t+i}, the growth rate in dividend (change in the natural log of dividend) between period $t + i - 1$ and $t + i$, is demeaned. The variable δ_t^* is a 'perfect foresight' or 'ex-post rational' log dividend–price ratio; it is the (log) dividend–price ratio that would obtain according to the present value model if everyone knew all future dividends. Although δ_t^* has a t subscript, it is not in fact known at time t and only its conditional expectation δ_t is observed. This model can be interpreted as saying that (log) dividend–price ratios are high when dividends are expected to decline over the future relevant to present values, and low when dividends are expected to rise. The parameter ρ is the discount factor, determined by the point chosen for the linearization. Campbell and Shiller (1988; 1989) set this equal $\exp(-(s - g))$ where s is the discount rate (computed as the average real return of the market over the historical sample) and g is the growth rate of real dividends (computed as the average growth rate over the historical sample). The assumption here is that the discount factor ρ is constant through time, though in fact variations in ρ may well induce more volatility in asset prices. Associated with this approximation is a linear approximation ξ_t for the (demeaned) one-period holding return between time t and time $t + 1$:

$$\xi_t = \delta_t - \rho\,\delta_{t+1} + \Delta d_t\,. \tag{4.4}$$

It follows from these definitions that:

$$\delta_t - \delta_t^* = \sum_{j=0}^{\infty} \rho^j \xi_{t+j} . \tag{4.5}$$

This means that the difference between the log dividend–price ratio today and the ex-post-rational log dividend–price ratio is the present value of (demeaned) holding period yields. Moreover, since holding period returns are by this model serially uncorrelated, it follows that the variance of the sum on the right hand side of (4.5) is the same as the sum of the variances. Therefore,

$$\text{var}(\xi_t) = (1 - \rho^2)\text{var}(\delta_t - \delta_t^*) . \tag{4.6}$$

Thus, the variance of returns is proportional to the variance of the unforecastable component of δ_t^*. Since we do not have a market for claims on aggregate income, we do not observe the market forecast δ_t of δ_t^*, and we cannot form the variance of the right-hand side of the above expression. If we are to make an estimate of the variance of ξ_t, then we must specify an information set and a forecasting model for dividends.

To understand the issues here, it may be helpful to note that one could compute a time series of values of δ_t^* using a time series of actual dividend data, if one were willing to make some assumption about values of Δd_t for the periods beyond the near end of our time series; with sufficiently long time series this procedure might work well enough in producing an approximation to δ_t^*. If one then computed the variance of this constructed δ_t^* with a sample that ends substantially before the end of our time series data, then one would have a rough indication of the variability of the true δ_t^*. But to get an estimate of the standard deviation of returns ξ_t, one must gain an impression how much of the variance of δ_t^* is forecastable. One could estimate the variance of ξ_t by doing a time series regression with this constructed δ_t^* as the dependent variable and a set of information variables available at time t as independent variables. If one assumes that the market forms expectations linearly in terms of this information set, then the estimated variance of the error term in the regression can be used in (4.6) for the variance of $\delta_t - \delta_t^*$, to provide an estimate of the variance of ξ_t.

Ultimately it will be difficult to estimate how much of the variability of δ_t^* is forecastable. The reason is that the series is dominated by long-term or low-frequency movements. There are not a lot of independent observations of δ_t^*, not a lot of effective degrees of freedom. At the same time, there are many economic variables that have a strong low-frequency component. There are too many candidate variables that might be used to forecast δ_t^* relative to the number of effective degrees of freedom. We do know, however, that since δ_t is the expectation conditional on publicly available information at time t of δ_t, the variance of δ_t^* is an upper bound on the variance of $\delta_t - \delta_t^*$ and hence $(1 - \rho^2)$ times this variance is an upper bound to the variance of ξ_t.

Of course, we can never be sure whether there might be other variables not included in our information set that would help us to forecast δ_t^*, and thereby reduce the variance of $\delta_t - \delta_t^*$. If we consider that the public may have superior information, then our estimated variance may be regarded as a sort of upper bound (this is the West inequality, 1988*b*).

Despite the essential ambiguity of any measure of the forecastability of δ_t, it is worthwhile pursuing at least whether the smoothness observed in short-run movements in Δd_t is suggestive that δ_t^* is substantially forecastable. For this, and to avoid having to use the approximation implicit in use of a constructed δ_t^*, we may use the time series model developed by Campbell and Shiller (1988; 1989). Consider the first-order vector autoregressive model $z_t = Az_{t-1} + u_t$ where the k-element vector z_t has first element Δd_{t-1}, and has other elements that are other information variables available at time t, and where the k-element vector u_t is a vector error term with (degenerate) variance matrix Ω. Since it is presumed here that the price p_t is the price at the beginning of year t, lagged dividends, but not year t dividends, are in the information set on which the price p_t can be based. Here, the other information variables included in z_t will be Δd_{t-2}, Δd_{t-3}, etc., in which case the first-order autoregressive model for z_t is really a higher-order scalar autoregressive model for Δd_t in companion form, and the elements of u_t after the first are all zero. Following Campbell and Shiller, since $E_t z_{t+k} = A^k z_t$, it follows that:

$$\delta_t = -e1'A(I - \rho A)^{-1} z_t \qquad (4.7)$$

where $e1$ is a k-element vector whose first element is 1, and the others zero. It also follows that ξ_t is proportional to the innovation u_{t+1} in z_{t+1}:[17]

$$\xi_t = e1'(I - \rho A)^{-1} u_{t+1} \tag{4.8}$$

and hence:

$$\text{var}(\xi_t) = e1'(I - \rho A)^{-1} \Omega \, (I - \rho A)^{-1'} e1 \,. \tag{4.9}$$

We can therefore estimate the time series model for z_t and thereby estimate the variance of ξ_t under the assumption that the market uses this model to forecast. In the sense defined by West (1988*b*), this is also an upper bound to the variance of ξ_t, since the market may have more information to forecast.

As a check on this method of estimating variances of returns from dividend data, I first estimated the implied variance of Standard and Poor's stock returns using data from 1871 to 1992 on real Standard and Poor dividends, updated from data in Shiller (1989), to see how close the estimated variance of returns corresponds to actual variance.[18] Past experience with methods analogous to those described here applied to such stock market data show that the estimated variance of ξ_t tend to be lower than the variance of actual returns; this is the excess volatility result noted many times (see Shiller, 1989 or West, 1988*a*) for a survey). Still, such methods do suggest substantial volatility for the stock market.[19] Here, the matrix A was estimated by regressing the vector z_t on z_{t-1} with ordinary least squares, where z_t has k elements, Δd_{t-j}, $j = 1, ..., k$. The value of ρ (the discount factor) assumed was 0.936, the same value that was used for the US stock market in Campbell and Shiller (1988; 1989). When the autoregressive model for real dividends was estimated with one lag, $k = 1$, the stock return volatility implied from equation (4.9) was 14.67%, but as k was increased the implied volatility steadily declined. With $k = 3$ the standard deviation of ξ_t was 12.13%, with $k = 5$ it was 10.90%, with $k = 10$ it was 10.47%, with $k = 15$ it was 8.14%, with $k = 20$ it was 7.00%, with $k = 30$ it was 5.29%.[20] Over the same sample period for which these returns were calculated in the $k = 10$ case (1882–1991), actual returns computed from these data had a standard deviation of 18.13%, higher than any of the standard deviations of our theoretical returns. Over this sample period the correlation between actual and theoretical returns was 0.36. This is not a smashing success for the expected present value model in predicting returns, especially considering that the correlation between the growth rate of dividends and stock market returns was 0.94 over this same sample.[21] The tendency of the standard deviation of theoretical returns to decline with k apparently reflects the finding of Campbell and Shiller (1988; 1989) that dividends relative to long moving averages of dividends help forecast dividends: when log real dividends are low relative to a long moving average of log real dividends, log real dividends tend to increase; when they are low, they tend to decrease. This behavior of dividends reflects a tendency for dividend movements away from the average dividends over recent history to be reversed. This behavior is akin to mean reversion of dividends, where the mean to which it reverts evolves slowly through time. The regression coefficients in the high k instances tend to be pretty consistently negative and, very roughly speaking, to die out exponentially with lag. An exponentially weighted distributed lag on past growth rates is, of course, proportional to the latest log dividend minus an exponentially weighted distributed lag on log dividends; i.e. proportional to the current log dividend minus a long average of lagged log dividends.

While this preliminary check on the method using stock market data suggests that the method is not terribly accurate in predicting the standard deviation of returns, perhaps the method is good enough to give us a rough indication of the order of magnitude of the volatility we might observe in perpetual claims markets in national incomes.

5
Real Estate and Other Markets

There are other income factors, besides the aggregate national income and labor income factors, that contribute as much uncertainty to the incomes of individuals and organizations as do many risks currently traded in financial markets. If those who retail insurance policies against risks of changes in values of claims on incomes or service flows are to be able to tailor their insurance to the various exposures that their different clients have to these risks, they will want to lay off in hedging markets the risks of changes in these income factors that influence them because they are providing the insurance policies. This chapter indicates some of the most salient of these other markets.

Several such markets will be considered here—real estate, unincorporated business, privately held corporations, consumer and producer price index components, and art and collectibles—as well as some thoughts on how we might go about systematically looking for the major risk factors to incomes for which new markets would be most useful.

Real estate

There is a substantial industry devoted to insuring real estate against physical damage, but there remains virtually no way for people to insure or hedge the risk of declines in value of real estate holdings. Portfolios of real estate tend to be highly undiversified, concentrated in small regions, and so people and organizations are accepting substantial risk of local real estate price fluctuations. Those who lend to property owners are also vulnerable to such risks, as are holders of real estate mortgages and mortgage-backed securities, and private mortgage insurers.

In the United States in 1990 the value of residential real estate was, according to estimates, $6.1 trillion and of commercial and industrial real estate $2.7 trillion.[1] These estimates put the value of residential real estate at close to twice that of the entire stock market in the United States, of commercial real estate at close to the value of the entire stock market.[2] Of course, as was stressed in the last chapter, domestic wealth is only a fraction of the value of a nation, when the value of labor income is included. Still, these numbers indicate that real estate in the United States is far more important than many assets that are enthusiastically traded in our liquid financial markets. The absence of good liquid markets to hedge aggregate real estate risk is indeed striking.

Markets in residential real estate and urban land

Price changes in residential real estate markets have been on many occasions appalling. The data on median price of existing homes in US cities produced by the National Association of Realtors (here converted to real prices by deflating with the US consumer price index, CPI-U) reveal many booms and busts in the last twenty years.[3] The booms have shown sustained dramatic increases in real prices at different times in different cities: the annual average rate of increase was 9.3% in San Francisco 1976–80, 17.7% in Boston 1983–7, 16.9% in New York 1983–7, 10.2% in Washington DC 1986–8, 19.1% in San Francisco 1987–9, 21.2% in Honolulu 1987–90, and 22.3% in Seattle 1988–90. There have also been at the same time dramatic real estate busts at different times in different cities: the annual average rate of change in real prices was –14.6% in Houston 1985–7, –12.2% in Oklahoma City 1986–9, –10.1% in New York 1988–91, and –13.5% in Boston 1989–91. Note the lack of synchronization of the booms and busts; this indicates that any hedging markets in real estate should be defined geographically in areas much smaller than the United States.

Some real estate booms and busts outside the United States have been even bigger. In fact, the estimated appreciation of Japanese land value in 1987 was greater than the total Japanese gross domestic product for that year. The appreciation in Korean land value exceeded the gross domestic product in that country in each year from 1988 to 1991; the appreciation was over twice gross domestic product in 1989. In contrast, at no time since 1986 has price appreciation in the stock market exceeded 34% of gross domestic product in either country.[4] In both countries, there has been widespread concern about the redistribution of wealth caused by these land price movements, and the potential for further economic disruptions should recent declines in land prices turn into a debacle.

The costs that people have borne as a result of such price fluctuations are indeed distressing. Those who waited to buy a home find themselves priced out of the market because of a boom. Fearing a possible boom, many people buy houses that they do not want at present. Declining real estate prices wipe out the savings of many people who used their savings on a down-payment on a house. Declining real estate prices have caused bankruptcies of savings and loan associations, banks, and mortgage insurers in the United States. Booming real estate prices have caused overbuilding, producing a glut of real estate that contributes to a subsequent crash in prices. All of these hazards could have been insured against had markets in real estate prices, or perpetual markets in real estate incomes, been available.

The real estate booms and busts have both a rational and a psychological component. Karl Case and I (1988) documented the psychological component of some recent booms by sending out identical questionnaires at the same time to recent homebuyers in four different cities in May 1988. In the two California cities, Anaheim and San Francisco, there was a residential real estate boom going on at that time, with single-family home prices rising around 20% a year. At the same time, in Boston there was a post-boom situation in the single-family home market, with prices leveling off after having roughly doubled in the mid-1980s. (The end of the 1980s in Boston were followed by sharp declines in prices, although that was not known at the time of our survey.) At the same time in Milwaukee, there was not and had not been recently any sign of a boom or bust in real estate. By comparing these cities using identical questionnaires at the same time, we were able to take advantage of a natural experiment that enabled us to isolate factors associated with real estate booms. We found that there were sharply different expectations for future price changes across the four cities. We asked: 'On average over the next ten years, how much do you expect the value of your property to change each year?' The mean answer in Anaheim was 14.3%, in San Francisco 14.8%, in Boston 8.7%, and in Milwaukee 7.3%. The much higher expectations in the two California cities for sustained price increases over the succeeding ten years would hardly seem rational, given the past behavior of real estate prices, and given that California real estate prices were already at very high levels. When asked to fill in their reasons for their expectations, people were very inarticulate, never citing any quantitative or authoritative evidence, instead tending to cite clichés ('California is a great place to live, everyone wants to come here') or to react to vivid events or things one might see driving around the town.

Had these homeowners been hedged, and their neighborhood risk been borne by diversified investors, then there may never have been the vivid, attention-grabbing situation that induces a speculative boom or bust. Of course, had hedging markets been around they might also have attracted some professional speculators, who might have created booms or busts where none would have happened. But it seems altogether more likely that making the real estate markets liquid and available to professional investors would reduce the potential for local booms and busts, by taking a load off the shoulders of the relatively ill-informed individual homeowners and opening the market to an international community of investors who may not be influenced by the local fads.

Speculative booms and busts in residential real estate markets today are potentially more damaging than those in financial markets, in that the participants usually have much of their wealth concentrated in that local market and may be highly leveraged through their mortgages. Moreover, the vividness of the event of a real estate bust is so dramatic to them that the homeowners may react more strongly than do corporate managers to the perceived decline in their net worth. The rapid collapse in prices of houses in some transitional neighborhoods is often associated with an exodus of responsible homeowners, who are trying to get out before the collapse goes further; the effect of this is to accelerate the deterioration of the neighborhood quality.

Most homeowners will probably not deal directly in hedging markets, but will do so indirectly through intermediaries. Retail products that are attractive to homeowners might look like insurance policies; in fact they would be cash-settled put options on the individual homes. Such insurance policies (put options) might be attractive to homeowners who do not want to give up the upside potential for the prices of their homes, and only to insure against loss, and the products would rather more resemble insurance policies with which the homeowner is already familiar.

Homeowners would want most to see their individual home prices insured. There are problems, however, with developing policies that directly insure the price of a house. Appraisers may find it difficult to be objective enough in their appraisals to satisfy those who stand to gain or lose large sums in contract settlement depending on the appraisal. There is probably no observed market price for the home at the time of settlement, since people sell infrequently. And if there is an observed market price, this price may be biased: the homeowner could arrange, for example for a non-arm's-length sale to a relative at a low price. Or the homeowner might expend little energy to find a suitable buyer. Moreover, the homeowner might not make efforts to maintain the property properly, thinking that the insurance policy will cover any decline in value that is caused by poor maintenance. Such risks could be reduced somewhat by having a percentage deductible for the policy.

While policies based on the actual price of a given home are most effective in reducing price risk for the homeowner, in fact eliminating such risk, they may be inefficient because of the potential for abuse by homeowners and appraisers in the measurement of price. Thus, they may not be marketable to the public at an attractive premium.

Alternatives to such policies are policies that insure the homeowner, not against price declines of his or her own home, but against price declines of the entire region. For example, a homeowner in Evanston, Illinois, would buy an insurance policy that pays if the price of homes in the greater Chicago area declines. The change in prices could be measured for a certain area by a residential real estate price index, such as those discussed in Chapters 6, 7, and 8 below. Such policies do not provide any incentive for individuals to make non-arm's-length sales, nor do they reduce any incentives for them to find a good buyer for their house, or to maintain the house properly. Since the policies are thus relatively safe from abuse, they can be provided more cheaply to homeowners. On the other hand, the home insurance policies may be a poor hedge for homeowners: house prices in Evanston may not closely track those in the greater Chicago area. Prices in individual communities may reflect such things as the tax rates, quality of schools, and crime problems that are specific to those communities. If the insurance payout offered to a homeowner is not well correlated with their risks, that is, if the hedge is poor, there may be no demand at all for the insurance policy.

Probably the ideal form of home equity insurance would take the form of insuring the homeowner against price declines in his or her neighborhood and real estate type. There could be price indices for high-priced, middle-priced, and low-priced homes for each zip code, and the homeowner could be insured against the drop in the index that pertains to his or her own home. A regional insurer, who issues many policies in a region represented by a futures contract, could hedge the risk incurred from having issued the policies by selling in the regional real estate futures market.

A recent effort in Chicago to sell insurance against the component of an individual home's price decline that is due to neighborhood effects has shown some modest success in its first three years. The 'home equity assurance program', was created in 1990 by voter referendum. Owners who register with the program pay a one-time $125 appraisal fee, and thereafter have added to their tax bill a fee of $12 to $25 per year, depending on the home's appraised value. After five years, registrants who sell for less than appraised value will be reimbursed for any losses due to decline in neighborhood conditions. (The program has not existed for five years yet, and so the reimbursement has not been tested.) The program administrators will have the authority to judge whether the loss incurred by the seller is due to neighborhood decline or other reasons. Only a few per cent of eligible homeowners have paid the appraisal fee to enroll in the program. Probably the program has not received enough publicity or endorsements from opinion leaders. Moreover, homeowners must under the present plan themselves take the initiative and bear the expense of enrolling in the program. They would be more likely to enroll if the option were presented to them as part of the procedure they go through when getting a mortgage or insuring their house. Undoubtedly, the judgmental nature of the settlement process has inhibited the growth of such programs. Private firms, which might do a better job of marketing, will find it difficult to offer such home equity assurance until they can decide upon a more verifiable method of judging the loss.

Effects of hedging markets on cash market inefficiency

The residential real estate markets appear to be notably inefficient. There is sizable inertia in prices: price trends continue for months and years. That there is inertia in the US housing market has been documented by Case and Shiller (1989), Poterba (1991), and Kuo (1993). Evidence for inertia in the Tokyo housing market was found by Ito and Hirono (1993). This tendency for inertia is readily visible in plots of good residential real estate price indices, indices that are so constructed that they are not disrupted too much by variations in the mix of houses sold or by sampling error. Fig. 5.1 shows an example of such a price index for Los Angeles County, 1985–92. Note the smooth, rounded appearance of the time series plot.[5] There was no smoothing at all involved in the construction of this index; the smoothness seen in the figure is a reflection of smoothness in actual price trends. There are indeed some sizable (and statistically significant) changes in direction within a year or so, especially toward the end of the sample, but there are also long periods of really smooth changes in prices.

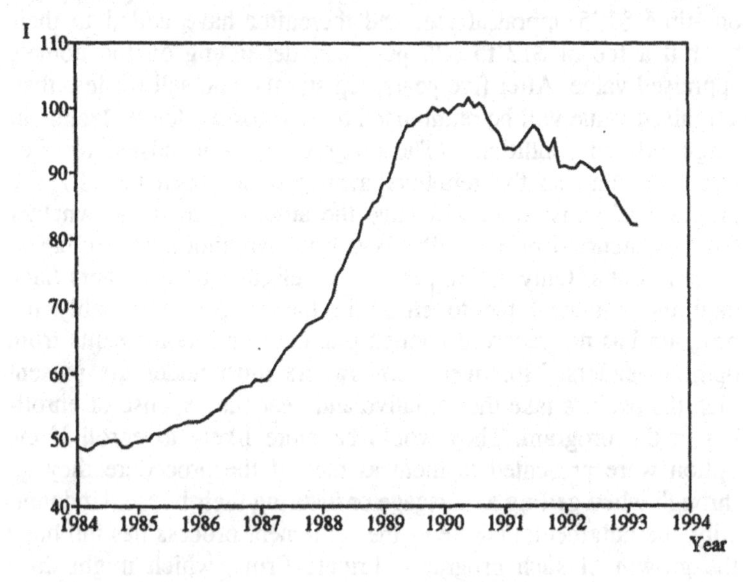

FIG. 5.1 Case–Shiller home price index, Los Angeles, monthly, 1984–1 to 1993–3.
Source: Case Shiller Weiss, Inc.

Strong inertia in price movements, of course, cannot persist in a liquid market; inertia as strong as that revealed in Fig. 5.1 would produce a powerful profit opportunity to those professional speculators who are accustomed to dealing with very tight profit margins. They would need only to buy when prices are rising strongly, sell when prices are declining strongly.

In illiquid markets such as this, trading to profit from serial correlation is difficult or impossible because of transactions costs. By definition, illiquid markets are markets where it is costly and difficult to trade. Brokerage costs of selling a house are often 6–8% of the value of the house, and the costs of storing a house that doesn't sell quickly can be far greater. Purchasing a house for investment runs the risk that one has bought an unrepresentative house, a lemon, a house not really demanded at the price except possibly by others with similar investment theories. Moreover, in illiquid markets there is not the price discovery seen in liquid markets; traders may not even know when prices have been increasing in recent months. This has certainly been the case with real estate markets, where price changes are very hard to discern from casual evidence. If speculators do not know what prices have done recently, then they cannot trade on that information, and cannot do their job of reducing inertia in price movements.

For contrast, a plot of the Standard and Poor's Composite Stock Price Index is shown for the same dates in Fig. 5.2. This plot shows more nearly the 'random walk' property of speculative prices that are well known to those who deal in these markets. While there would appear to be an uptrend in this series too, the trend here is not smooth, the serial correlation of price changes in this series is obviously dramatically lower than is the case with the residential real estate prices.[6]

To the extent that prices in liquid markets may be approximated by random walks, the variance of price changes will increase proportionally with the time interval over which the price change is measured. The variance of two-month price changes will be roughly twice the variance of the price change in one month, the variance of the price change in three months will be three times the variance of the price change in one month, etc. When prices are random walks, it does not matter over what interval one quotes the variance of price change; one can always convert this variance to that of a price change of another interval by multiplying by the

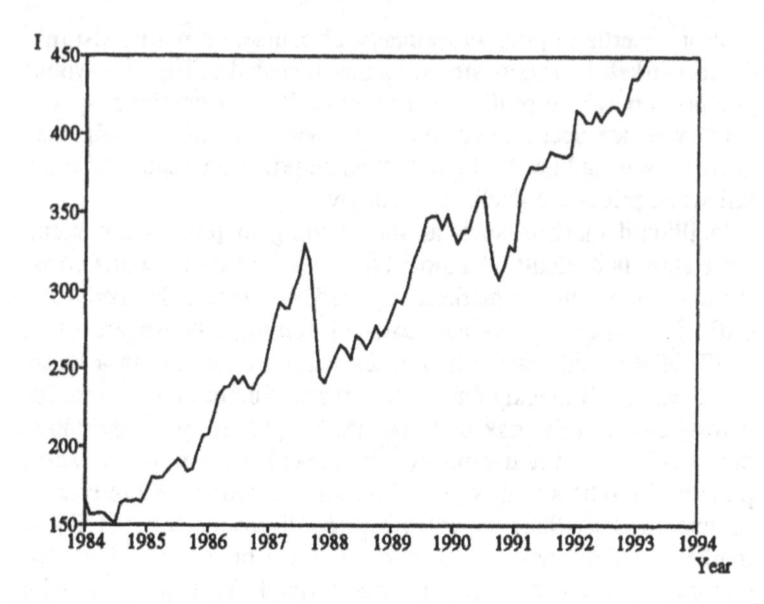

FIG. 5.2 Standard and Poor Composite Stock Price Index, monthly, 1984–1 to 1993–3.
Source: Standard and Poor's Statistical Service.

ratio of the lengths of the intervals. People accustomed to such prices are wont to consider a measure of price-change variance over one interval as good as over any other interval of time. This is not the case for the positively serially correlated price changes that we may often find in illiquid markets. For such price changes, the conditional variance may increase at an increasing rate with the horizon of the forecast. For such price series, it may matter very much over what interval we measure the price change when we compute variance. If we take small time intervals, then we will find that there is very little variance, and it might appear that there is little uncertainty to hedge; the appearance may be quite different if we pick a time interval to measure price changes.

This difference in the nature of uncertainty about prices also dictates a difference in the nature of the contracts that ought to be traded, and in the nature of the volume of trade. Longer horizon contracts, even perpetual claims or perpetual futures, ought to be traded with residential real estate futures, since there is little uncertainty about prices in the next few months. There is still substantial uncertainty about the values of the index a few years out.

Establishing hedging markets, such as futures or options markets, in housing is likely to have the effect of improving efficiency in the real estate market, making the cash market price look more like a random walk than it has in the past. A tendency may develop towards basis pricing in the housing market: asking prices of homes might be set in terms of futures market prices, so that the asking price adjusts automatically with the market. Sales agreements might also specify selling price in terms of futures market prices, reducing the risk that one of the parties will want to back out of the sales agreement if aggregate real estate prices change before the settlement date. Such practices would eliminate an important reason for inefficiency in real estate markets, making them immediately responsive to conditions in the liquid futures market.

If we disregard speculative effects, we might expect that establishing hedging markets in housing would also have the effect of increasing the price of housing, especially in the short run. If people can hedge their risks, then they will be more eager to hold housing. Putting it another way, the discount rate relevant to housing investments may fall, as investors see housing as part of a diversified portfolio.

In considering the long-run effect of the establishment of real estate hedging markets, we must consider the extent to which the supply of housing will respond to a price increase. If the values of properties priced as houses is due entirely to the land component, then there is no conceivable supply response; the long-run effect of establishing a hedging market will be only to increase prices. If housing were fully reproducible at a fixed cost, then the long-run impact of the establishment of hedging market would be to increase the quantity of housing, not its cost. In practice, the truth is somewhere in between these two extremes: the long-run effect is likely to be an increase in land prices and an increase in the supply of structures. The price of a typical house (lot and structure) will be somewhat higher; but the structure will also be somewhat larger than if there were no house hedging market. The rise in price of land should not be viewed as a social cost; quite the contrary, it is an increase in national wealth created by the improved risk management allowed by the establishment of the futures markets.

Of course, it is difficult to be assured that the effect of speculative pressures on real estate markets might not offset these general tendencies. The establishment of liquid markets brings in a whole new class of investors, whose attitudes matter for equilibrium in this market. The establishment of hedging markets might lower the price of real estate in overpriced markets, if it facilitates short-selling of real estate.[7]

Commercial real estate

Fluctuations in commercial real estate prices and rents cause the same sort of dislocations that we see in residential real estate markets. There are periodic office-building booms and gluts. Warehouse facilities are overbuilt in response to price increases, and then later left standing vacant. Corporations deciding to locate in one city may find that the cost of real estate services to them and to their employees have turned against them. Fluctuations in apartment rents have caused such turmoil that city governments have in many places imposed rent controls, setting ceilings on rents charged. People (or organizations or cities) could hedge themselves (or their employees, clients, or inhabitants) against the risks of changes in real estate prices or rents if there were liquid hedging markets.

Commercial real estate may deserve markets separate from residential real estate, since the substitution between commercial and residential real estate is not perfect, so that their prices have substantially different movements. Commercial real estate probably also shows rather dramatically different price movement across different real estate classes: apartment buildings, retail space, office space, factories, and warehouses. Of course, commercial real estate risk-management contracts would have to be defined for geographical regions smaller than the United States, because of interregional price change differences much like those observed with residential real estate.

Commercial real estate, however, provides some greater challenges, relative to residential real estate, to those who would define new markets. It is difficult to derive price indices of commercial real estate, since there are far fewer properties sold than is the case with residential real estate. Moreover, with few sales, manipulation of the futures market via sales in the cash market is easier.

Commercial properties tend to be altered more fundamentally between sales, and financing arrangements and other property transferred with the sale may complicate the sales transaction.

Despite such difficulties, there does appear to be a nascent market in index-based real estate swaps. In January 1993, a $20 million real estate swap deal was completed by Morgan Stanley & Co. Inc. and Aldrich, Eastman and Waltch L.P., a deal touted as the first real estate swap ever. One counterparty, who agreed to pay Morgan Stanley a return based on the Russell/NCREIF (Frank Russell National Council of Real Estate Investment Fiduciaries) index of commercial real estate prices and received LIBOR (London Interbank Offered Rate), a short-term interest rate, was a corporate pension fund. The other counterparty, who paid the interest rate and received the real estate return, was a life insurance company.

The proliferation of such risk-sharing arrangements could be enhanced if better measures, other than the appraisal-based Russell/NCREIF index, could be devised to settle such contracts. The judgmental nature of the index may certainly raise problems of objectivity. Moreover, there may be no way for appraisers to discharge their obligations to value real estate in an objective way, save developing their own transaction-based real estate price indices. Possibly repeat-sales indices, like those to be discussed below in Chapters 6, 7, and 8, could be devised for commercial real estate.

One way of developing an index of prices for apartment buildings is to use an index of condominium prices. When apartment buildings can be freely subdivided into condominium buildings, and when new construction for both is going on, there should be a close correlation between condominium prices and apartment prices. This would make it easier to get accurate indices for cash settlement, since there are many more sales of condominiums than of apartment buildings. Condominium sales may be regarded as sales of apartment buildings in pieces, though in fact there can be sizable spreads between prices of apartment buildings and the cumulated values of condominiums in comparable buildings.

Because of the infrequency of clean repeat-sales pairs of relatively unchanged properties, it may be better instead to devise a rental index for commercial properties, and to use a perpetual claims or perpetual futures market to discover the present value of the income flow. One would need only to collect data, by questionnaire survey, say, of the rents that are currently being paid on the commercial properties. Now, some indices of commercial rents attempt to collect data on newly negotiated rents, to make their index more forward-looking. Commercial rents are usually negotiated in multi-year contracts, and so their behavior is very sluggish. This method of index construction would not be desirable for the purpose of creating an index to use for cash settlement of a perpetual claims or perpetual futures contract. The market pricing will take care of the looking forward; it is essential rather to create an index of actual rents paid, no matter how sluggish this index may be.

Land containing natural resources

Oil is a storable commodity, and so it would seem that it is a natural candidate for a conventional futures market. Those hedging stocks of oil may well want a futures market that cash settles on the price of oil, or that involves physical delivery of oil. Moreover, since ownership of oil-producing land is in effect a claim on stored oil, it would seem that the same futures market should suffice for hedging the price of this land.

There are reasons, however, to suppose that oil prices might better be regarded as measures of the dividend on land, so that a futures market for the oil-producing land might better be a perpetual one.

There is, first of all, a cost to producing oil and to storing it. Oil below ground and oil above ground are not interchangeable. In times of extremely high oil prices, the cost to producing oil may rise precipitously, as the existing facilities for the extraction, transportation, and processing of oil are strained. There may be temporary shortages of oil, when stores of oil are depleted. Then the price of oil may well depart from the usual approximate random-walk property of storable assets, and there may be an expectation that oil prices will fall. There can be no expectation that oil-producing land prices will fall as much.

Moreover, production of natural resources such as oil do not seem to behave as one would expect if the stores underground were equivalent to stores above ground. So long as there are adequate stores of gold above ground to meet production uses, there is no point in owners of gold land extracting their gold, thereby incurring both production and storage costs. Brennan (1990) gave a couple of reasons why such gold production continues. One is that owners of gold-producing land fear government expropriation or other sanctions if the gold is not extracted. Another is that the gold must be extracted in order to signal to the market that it is in the ground. Natural resource prices like oil prices may be temporarily changed up or down because of the behavior of cartels, like the Organization of Petroleum Exporting Countries (OPEC).

It is possible that one could start a futures market in the price of oil-producing land. This might be important, since the value of oil-producing land is vastly higher than the value of the oil in storage tanks. Of course, oil-producing land is different from other land in that its value is likely to be determined by the expectations for the world price of oil. Idiosyncratic regional price movements are less likely. Yet oil-producing land is not inherently a liquid asset in the sense that others are, in that it is inherently difficult to prove the value of the oil it contains. Thus, an owner of oil-producing land may find it difficult to diversify his or her portfolio by selling off the land; it may be rather more easy to diversify by selling in an oil-land-price futures market.

A futures market in oil land prices could be set up based on observed prices of sales of oil-producing land, but this will be inherently difficult to do. The index would have to be based on repeat sales of the same properties (as will be discussed in Chapter 6 below) since the quality of the land (in terms of the type and amount of oil buried under it) cannot be verified and is likely to vary from plot to plot. But even a repeat sales index will suffer problems, since the value of a given plot of land may be diminished through time as the oil is depleted.

A perpetual claims or perpetual futures market on the price of oil may be a more natural choice to hedge the price of oil-producing land. We would expect prices in such markets not to respond to sudden spikes in oil prices that are perceived as temporary, rather to respond instead to information about the long-run outlook for oil prices.[8]

Unincorporated businesses and privately held corporations

Unincorporated businesses in the United States produce income that is at least of the same order of magnitude as that generated by incorporated businesses. According to the US National Income and Product Accounts, total non-farm proprietors' income was $325.4 billion in 1990, while total corporate profits before tax was $332.3 billion. In fact, however, it is better to think of cash flows generated by corporations as decidedly smaller than the cash flow generated by unincorporated businesses. Only corporations are subject to the corporate profits tax; profits after tax were $197.0 billion in 1990. Dividends paid were only $133.7 billion; in no year since these series begin in 1959 has dividend income reached half of proprietors' income.

Even the dividend income reported in the US National Income and Product Accounts exceeds the dividend income generated by publicly traded corporations; privately-held corporations generate dividends as well.

There are, of course, no large liquid markets on which claims on proprietors' income or on privately held corporations are traded. One can imagine such markets; there could be, say, perpetual claims or perpetual futures markets cash-settled on indices of their income.

The natural first place to look for indices to cash settle these contracts are in the national income and product accounts. We could set up, using these accounts, perpetual futures in proprietors' income and in dividends (which include the dividends of privately held corporations as well as public corporations). However, these data are possibly quite flawed for this purpose. These time series do not purport to measure the success of given businesses; as new businesses are founded there might be growth in these series even if there is no growth in the income of any one business. The often-noted growth of the service sector in the United States in the last couple of decades may have meant, for example, that many new unincorporated businesses have been established over that period.

To gain a crude sense of the potential variability of returns in markets for proprietors' income and for all dividends, it is still worthwhile to analyze these data series from the national income and product accounts. For this purpose, three time series were taken from the US national income and product accounts, annual data from 1959 to 1990: proprietors' income with inventory valuation and capital consumption adjustments, corporate dividends, and gross domestic product. All three of these series were deflated into real terms by the gross domestic product deflator and put on a per capita basis by dividing by the population of the United States.

These three series show substantial independence from each other. The correlation between five-year growth rates of real per capita proprietors' income and five-year growth rates of real per capita dividends is only 0.354. The correlation between five-year growth rates of real per capita proprietors' income and five-year growth rates of real per capita earnings is only 0.424. On the other hand, correlation between five-year growth rates of real per capita proprietors' income and five-year growth rates of real per capita gross domestic product is 0.571, suggesting that a substantial part of aggregate proprietors' income risk could be hedged on a gross domestic product futures market.

The method developed in the appendix to the last chapter of estimating standard deviations of returns in hypothetical perpetual futures markets was applied here to these real per capita data, with the same assumed value for the real discount factor ρ. Autoregressions with ten lags were estimated for each of these time series, and the standard deviation of theoretical returns in perpetual futures markets were inferred for each. It should be noted, by way of noting consistency of results, that the results roughly confirm, with this different data set and sample period, the results in the last chapter (using the Summers and Heston data) that the standard deviation of returns in a perpetual futures market for US gross domestic product is small: the standard deviation was here estimated (using US National Income and Product Account data) to be only 1.53% per year, which is fairly close to the value for the United States perpetual futures market in gross domestic product estimated in the preceding chapter. The standard deviation of the return in the market for proprietors' income was also small: only 1.68% per year. While this is a fairly small number, it is important to note that the theoretical return in the proprietors' income market has very little correlation with that in the gross domestic product market, only 0.37, indicating that what risk there is in this market cannot be readily cross hedged in a gross domestic product perpetual futures market. It is also quite possible, as noted in the previous chapter, that these estimates of the standard deviations of returns in these markets are on the low side. The same methods with these data produced an estimated standard deviation of annual theoretical returns in a market for claims on corporate dividends of only 5.66% per year, far below the actual standard deviation of annual returns in our stock markets. The correlation coefficient between the theoretical returns in the perpetual futures market in proprietors' income and the theoretical returns in the perpetual futures market for dividends is only 0.45, suggesting little ability to cross-hedge proprietors' income risk in a market for claims on corporate dividends.

Of course, the results with these postwar US national income and product accounts data are hardly definitive as to the potential variability of returns in markets for proprietors' income and for total dividends. Note that in the preceding chapter the variability of returns in perpetual claims markets for US gross domestic product was at the extreme low end of the variability estimated for countries around the world; the US economy has been very fortunate in this period in its absence of unpredicted major upheavals. The future variability of proprietors' income may indeed be much higher. Moreover, there may be much more variability of returns in perpetual futures markets based on indices of income of existing firms in specific lines of business. Given the size of the market for proprietors' income, there could well be room for many perpetual futures markets, distinguished by line of business.

Consumer and producer price-index futures

The consumer and producer price indices produced by governments suggest a number of possibly important new risk markets. Consumer price indices represent the inverse of the real cash flow on existing nominally defined bonds. They also represent the inverse of the real cash flow on any contracts with specified nominal payments, such as leases and labor contracts.

Moreover, many components of the consumer and producer price index published by government statistical agencies represent prices of investments in durable assets, prices whose changes through time represent risks to the owners of these assets. Examples of components include railway equipment, aircraft equipment, and ships.

Aggregate price-level risk

There is currently no true futures or options market in the world today where the risk of inflation, the risk of changes in the purchasing power of money, as measured by changes in the consumer price index (CPI), can be hedged directly. There do exist in some countries' index-linked bonds, and a portfolio of such bonds is automatically hedged against inflation. Index-linked bonds can also be used to hedge other inflation risks: one can form a portfolio that is short in index-linked bonds and long in non-indexed bonds: this portfolio will be a pure inflation portfolio. However, such markets, because of the difficulties in shorting, the lack of complete matching of coupons and maturities of both index-linked and non-indexed bonds, and the lower liquidity of government bonds, are not a substitute for true futures markets in the price level.

Ideally, it would seem that virtually all intertemporal contracts, including futures and options contracts, should be written in terms of *real* rather than nominal quantities, that is, the contracts should be indexed to inflation. People are ultimately concerned with their purchasing power rather than their holdings of money; contracts should be written in terms of their real concerns. Money is only the medium of exchange; it has no significance by itself. If the real value of the quantity of money changes dramatically through time, then nominal contracts introduce noise into people's wealth.

Most government bonds have been denominated in nominal rather than real terms. As a result, long-term increases in the rate of inflation have virtually wiped out many people who, for example, have put their retirement savings in long-term bonds. It is hard to see any potential offsetting advantage to these people in tying the real value of their retirement income to the price level by buying nominal bonds. If the price level were somehow correlated with other variables affecting their well-being, then it is conceivable that they might want to have invested in nominal, rather than real, bonds. But it is hard to make any case that tying the real value of bonds to the price level serves any such insurance purpose.

By the same reasoning, denominating long-term commodity futures contracts in nominal terms would also appear to be suboptimal. Consider a grain merchant storing grain for later sale. Suppose that the merchant hedges this by selling in a grain futures market. Suppose also that there is a sudden unexpected and dramatic increase in the aggregate price level as measured by the CPI. On average, we should expect that the price of the grain held in storage also increases at about the same rate. The real value of the grain is unchanged, and yet the merchant faces an adverse settlement on his grain futures contract. By selling grain futures, the merchant locked in, in effect, the *nominal* value of his grain, not the real value. If the inflation were really dramatic, then the merchant who hedged in the grain futures market might be wiped out, whereas nothing real would have happened had the merchant not hedged. Clearly, the grain futures market would better serve the merchant if it were denominated in real terms: that is, the price at which the merchant effectively sells his grain at expiration of the futures contract would be the contract price plus a correction for change in the consumer-price index or other national price index. By the same token, perpetual futures contracts might use an indexed return as the alternative asset in the settlement formula (3.1).

There is still a potential problem caused by indexing futures contracts against CPI inflation for some people who also have other existing contracts denominated in nominal terms. Suppose our grain merchant had borrowed at a fixed nominal long-term rate to purchase grain. Then hedging on a CPI-indexed futures market would not prevent an impact of inflation, since the inflation is still affecting the real value of the debt. If the grain is hedged on an indexed futures market, then the sudden inflation provides a windfall to this merchant: the real value of the grain is unchanged while the real value of the debt is wiped out. On the other side of this, if there were less, rather than more, inflation than expected, the merchant could be harmed, as the real value of the debt turns out to be more than expected while the real value of the grain-cum-futures contract is unchanged. This problem might be described as not inherent to the futures contract *per se*, but to the introduction of inflation risk through borrowing at the nominal interest rate. Had the merchant borrowed with a real interest rate, that is, by issuing a CPI-indexed bond, then the real value of the merchant's position would be protected by selling the grain in a CPI-indexed, rather than nominal, grain futures market. Or, had the merchant borrowed with very short nominal debt, rolling over the short-debt contracts as time proceeds into new short debt, then the merchant may be relatively insulated from the effects of inflation on debt, even though the debt is technically in nominal terms. The reason is that, to the extent that inflation is known up to the horizon of each debt contract, the market may tend to make the real interest rate constant from contract to contract, thereby making the sequence of loans effectively close to a CPI-indexed sequence. To the extent that the market behaves this way, that merchant may then wish to hedge in an indexed, rather than nominal, grain futures market. In practice, the tendency for short debt to preserve real interest rates in the face of inflation uncertainty is somewhat limited.

Index arbitrage would also be affected by denominating commodity futures contracts in real terms if there are not indexed debt markets for arbitrageurs. In order to arbitrage a CPI-indexed grain futures market whose futures price is too high relative to the grain cash market, the arbitrageur would buy the grain in the cash market and sell it in the CPI-indexed grain futures market, thereby locking in a real profit. But this real profit is not risk-free, however, if the arbitrageur must borrow at nominal rates to purchase the commodity for storage, since the real interest rate is uncertain.

There has not been much clamor in the US for CPI-indexed commodity futures markets.[9] One reason for the lack of interest may be a natural human tendency towards 'money illusion', a tendency for people to accept currency as a standard of measurement even when its value changes through time.[10] Such a human tendency is not immutable, and experience has shown that in times of major changes in currency value people do learn to adopt better standards. The proximate cause of the lack of interest in hedging inflation in the United States may be that inflation has been substantially forecastable over the horizons of most futures contracts. The inflation rate has been one of the most easily forecasted economic variables for macroeconomists; inflation rates are substantially serially correlated. To the extent that inflation rates are not serially correlated, the lack of perfect serial correlation may reflect to some extent noise in the price-collection process, and not genuine changes in underlying inflation.[11]

If the underlying inflation rate is much more forecastable than the price of the commodity represented by a commodity futures market, over the horizon of the futures contract, it is nearly irrelevant whether futures are in nominal or indexed terms. If an economic variable (here, inflation) that is known with virtual certainty out to the horizon of the futures contract is allowed to affect the settlement of the futures contract, it will have no effect on the contract except to cause the futures price to be scaled up or down by this effect. The effect of inflation on the futures price should have virtually no effect on the settlement, and hence no substantive effect at all.

For most commodities, over the horizons that have dominated futures trading, there is much more uncertainty about the commodity price than about the consumer price index. Hence, any discussion about the choice between nominal and real contracts is rather academic. On the other hand, if we either extend the horizon of the contract or adopt perpetual futures (so that inflation uncertainty over the life of the contract becomes more important), or if we choose a commodity whose price is less uncertain over the life of the contract (and so choose a settlement formula that is relatively sensitive to the commodity price at settlement time, so that the contract itself will represent substantial uncertainty), then the issue of real versus nominal contracts begins to assume more potential importance.

Consideration of indexing may be particularly important for hedging markets in assets whose cash markets are illiquid because the price in such markets may be substantially forecastable over short time intervals. Thus, the contract may need to have a longer horizon before settlement. If a short horizon is kept, then the contract may be too much influenced by consumer price index uncertainty.

A consumer price index futures contract might best be established as a perpetual claims or perpetual futures market or other long-horizon futures market. Since there is little uncertainty about inflation in the near future, most uncertainty about inflation is of the long-horizon variety. A single perpetual futures market in the consumer price index could be used in conjunction with any other perpetual futures market to convert, in effect, the alternative asset from being a nominal rate to an indexed rate.[12]

Components of consumer or producer price indices

Separate risk markets for the components of consumer or producer price indices might be advantageous to people who have concentrated exposure to risk of prices in some of these components, and who might wish to hedge these risks. Examples of such components include aircraft, ships, railroad equipment, trucks, agricultural machinery, computers, transformers and power regulators, oil-field and gas-field machinery, mining machinery, and paper industry machinery. People who own some of these assets might well want to hedge their risks of a change in their value.

Unfortunately, the producer price indices that we now have for these components are probably not ideal for use in settlement of contracts. The price indices measure prices of newly produced commodities, not the price of investments in existing assets. New indices would ideally be produced for these assets. These indices might be constructed using repeated measures methods, as described below in Chapter 6 through 8. The data input for these indices would then be repeat-sales data on used assets. Alternatively, perpetual futures contracts could be set up based on indices of rents of existing assets.

Indices of price of such assets, or prices in perpetual futures contracts for them, might behave radically differently from the price indices that are produced for the producer price index. For example, the index values of some of these assets might show a sharp tendency to decline with time, reflecting depreciation as well as other forces. For the purpose of playing a role in the cash settlement of contracts, such a tendency for the indices to decline is not a problem. Indeed, allowing the indices to reflect depreciation may mean that the risk management contracts may allow investors to manage the risk of uncertain depreciation as well as true price risk.

Agriculture

The first futures markets created were markets for agricultural products; these markets are very well developed and liquid. Yet these markets do not properly serve risk management of the agricultural community; notably, farmers make little use of them. It is worthwhile first to look at the agricultural futures market, which might appear to offer farmers useful risk-management services; to learn why these markets are not more successful; and then to try to see how new markets could be established. We shall see that farmers seem to have a puzzling resistance to hedging, but that part of the resistance is probably due to their difficulty in knowing the cost of hedging and, to the absence of appropriate long-term hedging vehicles. This will bring us to a proposal for futures market in agricultural land prices or perpetual futures markets in agricultural commodity prices.

The use of agricultural futures by farmers

Hedging of crops by farmers at planting time is often described as the archetypal case of hedging; financial textbooks have long featured stories of a farmer selling at planting time a futures contract to expire after harvest time. The textbooks explain that by selling in the futures market, the farmer becomes subject to settlements at the futures market that are negatively correlated with the value of the crop sold, thereby reducing the risk of the farming operation. Given that futures markets exist, it would seem that there is no reason why farmers should combine farming with speculation in the price of their crop; they can concentrate their attention on their business, which is the production of crops, leaving the speculation in crops to others who can diversify the risk better, or who have the inclination to take speculative risks.

Agricultural futures markets have been widespread for over a century, and farmers have had plenty of time to learn how to use them. Individual farmers have a great personal need to hedge, since their income often depends substantially on a single crop. Lenders to farmers also have reason to try to encourage farmers to hedge their crops, since such hedging activity reduces the likelihood of default. Prices of agricultural commodities are highly variable, and there are periodic farm crises in which many farms become bankrupt.

Evidence that farmers have not been the main users of futures markets is provided by the fact that, usually, open interest in agricultural futures contracts has not tended to be high during the growing season. It was stressed long ago (Working, 1953) that open interest tends to rise and fall each year in rough correspondence with the volume of the commodity in commercial hands; this means that open interest rises dramatically after the harvest, not after the start of the growing season.

Helmuth (1977) undertook a survey of US farmers with annual gross commodity sales greater than $10,000 which revealed that in 1976 only 5.6% of farmers bought or sold futures for any commodity. The proportion who traded increased with the annual gross commodity sales, amounting to 1.0% of farmers with sales in the $10,000–19,999 range, and 13.1% of farmers in the greater than $100,000 range. But even for the large farming operations, only a small proportion traded in futures markets.

Most of these trades in futures markets were apparently for speculative rather than hedging purposes, as of those who bought or sold futures, only 33.4% bought or sold contracts for commodities used or produced in their own farming operations. (It should be noted that some farmers cross hedge, e.g. hedging sorghum or barley in corn futures.) And of those who bought and sold futures contracts for commodities they used in their own farming operations, 74.9% *bought* (and later sold) contracts; only 51.5% of these *sold* (and later bought) contracts. This means that their positions showed some tendency to be long, rather than short, and, Helmuth concluded, 'that for grain farmers this would imply a bias towards speculation', although for livestock farmers, who may be long hedging inputs, 'implications would not be as strong'.

Helmuth's survey asked farmers 'What are the reasons you have not bought or sold futures contracts?' They were given nine possible answers and an 'other' category to choose from. The most popular answer was 'not acquainted with how futures market operates', chosen by 28.0% of respondents. A substantial number choosing this answer is to be expected, given that there are so few farmers using the contracts; it is not really a reason not to hedge, but is the consequence of decisions not to hedge. The next most popular answer was 'size of farming operation was too small to warrant using futures contracts', chosen by 20.1%. It is true that contract size is larger than some farmers' individual crops, reflecting the lack of interest by many futures exchanges in interesting small farmers in trading. The next most popular answer, 'futures market too risky', chosen by 13.3% of respondents, reveals that many respondents assumed that the question was about speculation, rather than hedging, and well they might assume this, given the relative paucity of hedging use of these markets. None of the other answers garnered more than 10% of responses: only 9.8% chose 'lack of adequate capital' (presumably to put up margin for the futures contract), 9.2% chose 'don't have time to follow the futures market carefully', 6.2% chose 'don't approve of the futures market', 3.9% chose 'fear of being "locked in" by limit moves in futures prices', 2.8% chose 'the past year offered no opportunities worth trading', 1.9% chose 'my local basis (cash price—futures) is too unstable', and 4.8% chose 'other'. It is, of course, difficult to ascertain from short survey questionnaires the ultimate reason why farmers do not make more hedging use of futures markets.

That many farmers speculate in their own farm outputs is not necessarily contrary to rational behavior; farmers do have an advantage in information about the market for their own crops, due to their involvement in the production of that crop, and so it might be logical for them to speculate in that market rather than another market. Their doing this might be considered as useful, contributing their knowledge towards making futures markets more efficient. Still, rational risk management would seem to require that farmers should tend to be short their own products more often than long. There may be psychological tendencies that would encourage farmers to over-speculate in their own products, due to the vividness to them of price changes and their personal involvement in the market.

Hedging in agricultural markets has been done more by the merchants and grain millers, for which, one would think, the incentive to hedge is smaller. To the extent that these merchants and grain millers are corporations that are owned by investors with diversified portfolios, the actual risk is spread very thinly over many people, quite in contrast to the risk that an individual farmer faces with his crop.

One reason that individual farmers do not hedge may be just that the kind of sophisticated judgment about risks does not develop until the risks are very large, as in very large farming operations, where sophisticated expertise can be purchased. According to Gray (1977, p. 339):

> Futures markets grow only out of hedging needs, which in turn are perceived most clearly by specialists who have a large stake in prices. Merchants and flour millers, for example, have always used wheat futures almost without exception—bakers almost without exception have not. Cattle futures trading grew much more rapidly than hog futures because cattle feeding had grown more specialized and to a larger scale—hog futures trading began to grow as specialized hog enterprises emerged.

Gray argues that farm products that are hedged well are those that are grown by sophisticated large operators.

Intermediaries

Economic theory would suggest that, indeed, farmers might not themselves best hedge in the futures markets, but instead deal with intermediaries who in turn hedge for them in the futures markets. Farmers can make forward contracts at planting time to sell their crop to merchants at prearranged prices; the merchant can then hedge the risk that was acquired by the forward contract. A reason why this may be better than the farmer's own hedging in the futures market is that the merchant can diversify across farms, and therefore can insure the individual farmer against the local variations in price that account for basis risk. Moreover, by diversifying across farms the merchant can efficiently provide contract provisions that allow the farmer to deliver what his or her farm produces, rather than a pre-specified amount, thereby tailoring the contract size to the crop produced. Finally, the merchant, who may have personnel at the local agricultural sites, better knows the farmer's creditworthiness, and therefore might be able to eliminate the need for margin accounts and the associated margin calls. In short, the local personnel may well in effect provide a joint product of agricultural delivery point and provider of certain banking services. Some farmers have trouble coming up with margin and meeting the margin calls.

But the open-interest evidence tells us also that farmers are not, for the most part, indirectly hedging on the futures market. If most farmers were doing so, then we would expect to see very substantial open interest during the growing season.

There does, however, appear to be somewhat more use by farmers of forward contracting than of futures contracts. According to a survey by Heifner *et al.* (1977), 20% of corn purchases, 14% of wheat purchases, and 20% of soybean purchases from farmers in 1974 were forward contracts where price was settled more than 30 days before delivery. Though still only a fraction of all sales are hedged in these crops, we do see a substantial number of people whose hedging needs are being served by forward markets. And the purchasers from these farmers show a distinct tendency to hedge their own risks. According to a US Department of Agriculture Grain Industry Survey, reported in Helmuth (1977), 72.7% of subterminal elevators, 58.6% of terminal elevators, 67.2% of export elevators, 81.4% of soybean processors, and 41.7% of feedlot elevators hedge routinely. Only the country elevators show a low tendency to hedge: of these, only 16.5% hedge routinely.

Some of the selling of forward contracts on farmers' own crops is probably not actually hedging behavior, because the farmer undertakes the contract because of an *expected* price fall, rather than to hedge risk. Farmers who do sell their crop in advance on forward markets (as well as those who sell on futures markets) may be merely using this vehicle to speculate on the market, and to profit from their opinion that prices will fall. Indeed, as noted by Allen *et al.* (1977), forward sales at prearranged prices tend to occur at times when extraordinary demand suddenly pushes prices up at planting time, encouraging farmers to want to lock in their profits. They note that in 1973, when cotton prices were abnormally high at planting time, buyers were 'scurrying about to line up supplies', and 40% of the US cotton acreage had been sold in forward purchase agreements before the planting season; 75% of the acreage was contracted before the harvest (Allen *et al.*, 1977, p. 2). Of course, we cannot tell from this information alone whether the sudden willingness of farmers to sell their crop in forward contracts represents an expectation that prices will fall or a feeling that prices may be especially volatile at the time of abnormally high prices.

Alongside the farmers who use intermediaries to hedge, there are other farmers who appear by their behavior in the forward markets to be deliberately speculating in the crops they raise, taking on even more risk than they need to. Farmers who contract to sell their crops often agree on basis pricing for their crop: the agreement is that the price on delivery will be related by a formula to the price in the futures markets. The price will be the futures price minus an agreed amount called the basis. Farmers who make forward sales under such an arrangement are merely assuring themselves of a place to deliver their crop, while they could as well have agreed on the price in advance. They are thereby deliberately assuming risk that would have been natural to eliminate at the time the forward contract was signed. According to Heifner's survey, in 1974, 66.0% of terminal and subterminal elevators and 28.4% of country elevators 'routinely' used basis pricing.

Many farmers sell their crops with delayed pricing, an institution that is really the opposite of a forward sale: the farmer delivers and transfers title to his crop to a merchant, who promises to pay the farmer at a later date according to a formula (using basis pricing or the merchant's posted offer prices at the later date). Farmers who choose the delayed-pricing option are speculating after their natural connection with the commodity is over. According to Heifner, 8% of corn, 18% of wheat, and 11% of soybeans purchased from farmers in 1974 were subject to delayed pricing. Note that, with wheat, more of the farmers sold with delayed pricing than with price fixed over 30 days in advance.

This case study of farmer's hedging of their crops shows that time alone will not create success for hedging markets among all parties; ordinary farmers do not seem to be hedging properties over a hundred years after the introduction of the contracts. We have seen that it is understandable that many farmers do not hedge, given the difficulty in proving the advantage to hedging, and given their ability to hedge their risks partially by diversifying across crops.

The only modest success of such grain elevators in getting farmers to sign up for risk management services is probably due largely to the difficulty in proving to farmers the advantages of such services, and the weaknesses of the contracts in serving their hedging needs, to which we now turn.

Difficulties measuring the cost of hedging

An important reason why farmers may not hedge is that they may suspect that hedging is too costly to them relative to the risk-management benefits it confers. They may well wonder whether, given these costs, it serves them well to hedge. Helmuth's questionnaire about reasons for not hedging did not list this reason. In any event, farmers may not be able to articulate this reason, feeling only that they have not been convinced that hedging contracts are a good deal for them.

The transactions costs to hedging are easily described. However, the costs imposed by backwardation, the alleged tendency for short positions in futures markets to lose money on average—that is, the tendency of futures prices to rise on average through time—are not so easily measured, and farmers might well imagine that costs are higher than they are in fact.

Berck (1981, p. 473) claimed evidence of substantial backwardation in the cotton futures market, evidence that the futures price tends to rise with time, thereby imposing a cost (a risk premium or cost of insuring) on cotton farmers who hedge. He estimated, using data on planting-time hedges, that 'a short position in cotton loses 9.3 cents per pound or about half of the per pound profits of growing cotton', and concluded that most farmers might well want, given their abilities to reduce the price risk by diversifying production across crops, to forgo hedging because it is too expensive. However, Berck's evidence of backwardation was not statistically significant. Other analyses of agricultural futures prices do not find significant evidence of backwardation.

The problem in verifying to what extent backwardation occurs is that the noise in futures prices is so large next to any presumed tendency to rise with time that it is not possible to estimate the extent of the average rise accurately, even with advanced econometric techniques. There is no wonder that farmers do not know the extent of backwardation.

Putting this point in more direct terms, a farmer does not know easily whether hedging has been advantageous, even after years of experience with hedging. It is easy for the farmer to see that risk has been reduced by hedging, but it is hard to know what evidence this personal experience implies for the expected cost of reducing this risk. Comparing how much the farmer would have made on average had hedging been neglected with the amount actually made, the farmer will run into the same issue of statistical significance that plagues scholars trying to determine whether backwardation exists. The amount the farmer would have made is so noisy that one cannot estimate its mean well.

Economists often tend to think of the advantages of hedging in futures markets as being self evident, and consider that there is no need to estimate the extent of backwardation in order to establish the general principle that hedging tends to be advantageous. This opinion of economists derives from a kind of theory of hedging equilibria, where the extent of backwardation is essentially a market-clearing price for bearing of risk. Those on the long side of the futures contracts can diversify over many crops and other assets, which may show negligible correlation with this crop, and so they might be expected to demand little extra return for bearing this risk. But we cannot expect farmers as a group to understand and accept such theories. The best we could hope is that opinion leaders impress on them the general validity of such theories.

Inadequacy of short-term hedging to manage farmers' risk well

The archetypal story of farmer's use of hedging vehicles, mentioned above, has it that farmers are supposed to hedge each crop separately; this may be called annual hedging. With annual hedging, when the farmer plants, it is said, the farmer sells short contracts that hedge this crop's price risk. But the price risk that a farmer faces with a particular crop may not become totally evident during the growing season just prior to the crop, and so the hedging may not work well.

An alternative type of hedging that farmers might use with existing agricultural futures markets is sequential rollover hedging. A farmer, for example, who wishes to hedge the risk of the next n crops may, now, sell n times as many futures contracts as would be sold if the farmer were hedging only one crop. In the next year, $n - 1$ futures contracts would be sold, and so on, until, the last year, when only one contract is sold.

If we disregard transactions costs and basis risk (risk that the futures price will not converge to the cash price obtainable by the farmer on the settlement date), then annual hedging is optimal for any n-year horizon if crop price movements are independent from year to year, and information about price movements occur only during the growing season prior to the crop. Under these assumptions, rollover hedging is optimal if the price movements are a random walk. In the former case, annual hedging would eliminate all risk, while rollover hedging would introduce risk, by creating $n - 1$ too many settlements this period that are unrelated to the risks of subsequent crops. In the latter case, rollover hedging would eliminate all risk, but annual hedging would fail to hedge the price changes that are unfolding today in relation to future crops.

In reality, of course, neither of these extreme cases is likely to obtain: information about future cash prices is neither independent from crop year to crop year, nor a random walk. In this case, no hedging mechanism with existing markets can insulate from all risks; only multi-year futures contracts could do that (disregarding basis risk).

Because of the way transactions costs are calculated in existing markets, rollover hedging of long-term crop price risks is prohibitively expensive. To hedge *n* years of crops this year, the farmer would have to multiply transactions costs by *n*. Even if transactions costs were only 1% per contract year, this cost of hedging could easily wipe out profit margins.

But according to a study by Gardner (1989), the annual hedging that farmers can achieve without incurring prohibitive transactions costs does not eliminate much of the risk that a farmer faces. Gardner finds, using data from 1973 to 1985, that when annual hedging is applied by a farmer in central Illinois to soybeans (Chicago Board of Trade futures) over a three-year horizon, the standard deviation of unhedged risk for a three-year horizon is 59 cents per bushel, compared to 33 cents for sequential rollover hedging and 8 cents for multi-year futures. (The last figure, for the multi-year futures that do not exist, is just the basis risk in the existing market, based on the cash price in central Illinois.) Gardner's corresponding figures for corn (Chicago Board of Trade futures, central Illinois cash) were 44 cents, 15 cents, and 4 cents per bushel, and for cotton (New York Cotton Exchange futures, Memphis cash) were 11.2 cents, 3.7 cents, and 0.7 cents per pound. Of course, the gap between the unhedged risk of annual hedging and unhedged risk of multi-year hedging would be even greater if we chose a horizon of more than three years.

On this account, then, it is not surprising that most farmers do not hedge. It was noted above that purchasing insurance against risk is very attractive only if the insurance can be described as virtually eliminating risk. Here, annual hedging hardly makes a dent in risk.

It should be noted that farmers do have the option, if commodity prices should deteriorate, of switching to a different crop. When this option is operative, it would mean that there is less demand by farmers for multi-year futures. The significance of this option varies; some farmers have land, equipment, and labor force that are strongly suited to a particular crop, others are more free to switch.

Creating long-term hedging markets in agriculture

Longer-term hedging markets would serve many farmers better than existing markets. Perpetual futures in agricultural commodity prices would serve very well those farmers whose land or other investments commit them strongly to a particular crop. If the land, human capital, and equipment a farmer has is suitable to production of only a certain agricultural commodity, then the perpetual futures in that commodity price may well reduce to a futures market on the value of the farmer's investment. A farmer whose land, human capital, and equipment are suitable to production of two or three different commodities might find it useful to hedge in all the relevant futures markets. Of course, there is no reason why transactions costs should be as high for a perpetual futures contract or for other retail vehicles as they were for the sequential rollover futures described above.[13]

Futures markets in the value of their farms and/or equipment might serve farmers' needs much better than futures markets in individual crop prices, and retail markets that insure this, rather than their crop prices, may be of more interest to them. Farmers switch between crops, and hence individual crop prices may not correlate well with the long-term risks that individual farmers actually face by having invested in a farming operation. The discussion of Chapter 3 suggests two ways that such markets could be set up. These could be perpetual futures markets on farm income in a certain region, or they could be ordinary futures prices on farm prices in a region.

Agricultural land income might be measured more easily than can commercial or residential land income, improving the possibility of hedging its risk through the use of perpetual futures markets. A survey of after-tax farm incomes per acre could be used to construct a series that would function as the dividend in the cash settlement formula (3.1) for perpetual futures.

On the other hand, agricultural land prices might capture farm risks better than agricultural land income. The prices are sensitive, too, to potential land uses alternative to agriculture, and they are not sensitive to accounting conventions, such as depreciation methods. Farmland price indices could be derived from data on sales of farms just as residential real estate price indices can.[14] There may be somewhat more problems, though, in deriving agricultural land price indices than residential real estate price indices. There are fewer farms sold than there are houses sold, so the price indices might turn out to be more noisy. Moreover, farm land tends to be chopped up and sold in pieces more often than does land under other kinds of real estate. There has been a secular trend towards smaller farm size around urban centers, so that the component of farm price that is due to the house situated on it may have been growing.

Art and collectibles

Hedging markets in art and collectibles would appear to be rather less likely to succeed than would the other markets considered here, since the volume of trade in art is lower than in many financial markets, and since the qualities of the different items may be much more diverse than in conventional commodity markets. Still, it is interesting to consider such a less-promising market as a case study, and a tentative case for such markets someday may be made.

While a substantial part of the price risk of an individual work of art can be diversified merely by holding a large collection of art, by diversifying only across individual works of art, there is still substantial market risk in art collections. Goetzmann (1992), using data (from Reitlinger and Mayer) on repeat sales of important paintings from 1716 to 1977, estimated that the standard deviation of the change in (log) price of an individual painting held for five years is very high: 79.2% per year, making individual paintings like some of the most speculative stocks. This risk cannot be diversified away by investing in many paintings: the market component of the risk, the standard deviation of the (log) price change in all paintings in the sample, was estimated at 56.5% per year. This tendency for different paintings to show similar price movements is also apparent in the Sotheby's Art Index. In the boom years of the late 1980s, all categories of paintings (old master paintings, nineteenth-century European paintings, Impressionist art, modern paintings, and American paintings) appreciated rapidly. All of these categories peaked in 1989 or 1990; all of them dropped in value between 1990 and 1991.

Those who maintain collections of art and other valuables are inherently undiversified—one might say that the collector has his or her own private value for the art as part of a collection, and therefore cannot hold a diversified portfolio of art. Museums could rent or lease or borrow the items that they display from other more diversified portfolio investors, but this is not the usual practice. Museums often feel that, since putting an object on regular display enhances its value, they wish to own it or have a commitment to buy it before doing this. Presumably, many collectors, whether institutional or individual, are substantially in the business of trying to predict which pieces will be ultimately regarded as of high value, and thus they are speculators. Calling them this may put an unfortunately commercial sound to their activities, and certainly for many the pure aesthetic motive for collections transcends concerns about price appreciation; still, most collectors cannot but be aware of the speculative nature of their business, and there are certainly incentives for them to invest in pieces that they think will appreciate.

These same collectors may not wish to speculate on the entire art market. Just as with investors in any other speculative asset, the private information that they have that enables them to do well in their investments is not information about the path of the entire market, rather about specific pieces, artists, or genres. Thus they have an incentive to hedge their collection, to insulate it from the movements in the entire art market. If they are concerned about possible price declines, there is, of course, little option to sell temporarily: collections represent the most illiquid of investments, in the sense that a good collection must be painstakingly assembled over years of search, study, and other expense; the incentive is instead to hedge the market risk by going short in the art market. Those who take the other side of a futures contract, the longs, are able in effect to invest in the entire art market without getting involved in the process of choosing individual pieces of art. For them, the investment may be regarded as a matter of expanding the diversification of their financial portfolios.

In the absence of a hedging market for art,[15] collectors could try to cross hedge in existing futures or options markets. A likely candidate would be the stock index futures market. Goetzmann (1992) documented a consistently positive correlation between price changes in the market for paintings and the market for stocks. When he regressed annual inflation-adjusted art returns (based on his repeat-sales index of prices of paintings) on real London Stock Exchange capital appreciation 1720–1986, he found an R^2 of 0.30. (The beta was 1.39.) Yet this R^2 is not so high that the cross hedge will be highly effective in reducing price risk for art collectors.

One problem in establishing hedging markets in art is that major museum collections, by common agreement, rarely deaccession works of art, except to provide the funds to buy other works of art. The Association of Art Museum Directors sanctions those who are net sellers of art; for example, the Rose Art Gallery of Brandeis University was sanctioned in 1991 for selling part of its collection to meet operating expenses. If museums really intend never to cash in on the value of their collections, then the ups and downs of prices in the art market may be of no concern to them, and they may have little incentive to hedge.

Of course, these restrictions against net selling apply only to museums, and not to private collectors or the gallery industry. And even museums themselves *are* at times net sellers of art, and the potential to sell is certainly of value to them even if they reserve such sales for extreme situations or for the distant future. The art collection is a fall-back source of revenue, a source against which the institution could conceivably borrow. For example, those who lend to a university must certainly be aware of the potential value in the university's art collection. Thus a decline in the value of a collection does matter to the institution, and thus they might have an incentive to hedge their holdings.

However, given their resistance to selling works of art there may be a serious inherent problem in museums hedging in art futures or options markets: should the market move up, instead of down, the museum would be faced with making good the losses. This might sometimes force a museum to be a net seller of art.

Museums, with their restrictions against net selling, may be reluctant to take short positions in a futures market, even if the other side of the risk is that they will, in a down market, gain funds to buy more works of art and become net buyers.

Another problem in establishing futures or options markets in art and collectibles is that it may be difficult to create indices that reliably measure prices in these markets. Of course, the markets must rely on price indices; they would certainly be cash settled, based on indices of art prices. The problem of the unrepresentativeness of sales in these markets is sometimes severe: in the late 1980s, for example, there were many sales of Picassos at rapidly rising prices; more recently, major Picassos have not been much in evidence at the auctions. To the extent that Picassos' prices move similarly to the entire art market, then the representativeness problems can be dealt with with ordinary repeated-measures price index methods (see Chapter 6 below). Goetzmann (1992) created such art price indices. Still, any such methods have limited ability to take account of the fact that different kinds of art may have different price movements; these kinds of works may be hard to identify objectively.

The tendency of works of art to move through auctions may be more related to their characteristics than is the tendency of houses to sell. The normal turnover of houses due to job change and change in family structure is not present in the art market. Thus, in creating an index of art prices for cash settlement purposes, it may be especially important to take account of characteristics variables, as will be discussed in Chapters 6 and 7. It may be essential to take account of many characteristics, even down to the identity of the individual artist.

Now, indices of art prices may not be producible very frequently. There is a substantial seasonal component to activity in the art market, with major auctions occurring in the Fall and Spring. For this reason, Sotheby's has not been producing a regular monthly index; usually a sector of their index is updated three or four times a year. Probably, reliable indices may not be producible much more frequently than that. Now, of course, while it might suggest some seasonality to the volume of trade in futures and options markets, this seasonality poses no clear obstacles to the success of the markets. Supposing that contract months are chosen to coincide with (or closely to follow) the active periods in the market, no problem necessarily arises from the absence of an index in the intervening times. The index is absolutely necessary only on the date of cash settlement. Thus a semi-annual index would be adequate for a futures or options market that settles twice a year. It is possibly better that an index should be produced more frequently than that, based on the few intervening sales, even though the index may be very noisy for those dates. Certainly, those who participate in the market will be aware that the index values for periods outside the usual seasons for art sales are unreliable, and so no harm is done in publishing such index values, assuming that no contract settles on the intervening index values. Ultimately, the prices in the futures or options market would provide the most timely information about market prices in art and collectibles.

Before starting trading on any art or collectible futures or options market, great care would be needed in evaluating the price index that is the basis of cash settlement. The Sotheby's Art Index, the only major published art index, may not be adequate for this purpose. The index is not strictly based on actual sales. For each of the thirteen sectors, say old master paintings, there is a fixed market basket of individual works of art, on average 30–40 works. Whenever a particular sector is updated, the expert responsible for that sector subjectively assigns a sales price to each work in the basket for that sector, an estimate of what that object would sell for then. But what is the expert to do in appraising a work that has not sold recently? The year-to-year change in the index must reflect a lot of guesswork. The Sotheby's index would appear to have even greater potential problems than the appraisal-based indices of commercial real estate, since it is so much more judgmental to identify recently sold works of art that are comparable than it is to find real estate properties that are comparable.

Systematic approaches to finding other markets

We have seen in this chapter a number of possible new hedging and insurance markets. The most important of these were the markets that allowed hedging of major components of individuals' and organizations' incomes including service flows.

A different approach to identifying potential new markets could take the form of modelling the tendency for comovement of incomes, systematically to identify the most important components. One could imagine doing some factor-analytic modelling to discover factors underlying variation in prices of claims on incomes, to enable contracts to cash settle on the basis of these factors. Doing this would be analogous to the factor analysis of security returns of Ross (1976) and Chen *et al.* (1986), but without data on returns. The conditional variance matrix of returns would have to be inferred from study of the time series properties of incomes.

One could obtain time series data on incomes in many narrowly defined occupations, or on incomes in many small geographical regions, or on incomes of many narrowly defined kinds of firms. Or one could obtain data on service flows (or prices) of real estate of many narrowly defined kinds, or in tiny geographical areas.

The technology available for inferring the underlying factors, broadly defined, is quite rich. After having derived proxies for returns on perpetual claims on incomes, factor analysis could be used, which would identify factors as linear combinations of the returns observed. There are observable factor models, and MIMIC (multiple indicator, multiple cause model) models, that would estimate factors as functions of other variables. There are geographical contour estimation models and clustering models that would allow regions or areas to be defined for which indexes could be used to settle contracts.

In practice, the idea of such modelling would have to be pursued with care, since the methods might produce factors that have no simple intuitive basis, and there may be skepticism that an estimated factor structure will continue to hold up in the indefinite future. A less formal approach to identifying factors on which to base new markets, an approach such as that taken earlier in this chapter, may suffice.

6
The Construction of Index Numbers for Contract Settlement

When creating indices intended for use in cash settlement of futures contracts (or perpetual claims or options, or swaps, or other over-the-counter forward contracts or retail insurance contracts), it is critical that we make each index represent value associated with a standard claim on future income (or services). We want our contract settlement to reflect the price of claims on income streams, so that the market can be used to hedge the risk associated with the claims. The problem is that our observations on prices or incomes may apply to dissimilar claims.

If we are making an index of housing prices (prices of claims on future housing services), we want to take account of the fact that at some times better houses may be sold than at others, so that the change in average price of a house reflects the changing composition of the houses sold rather than the change in price of a house. There are difficulties in taking account of quality, since we do not have definitive, objective measures of the quality of houses. If we are making an index of incomes in a particular region or occupation, we want to make sure that changes in the index do not reflect changing demographics (changing birth rates result in more young people in the area, let us say) or inflow or outflow of people to or from that region or occupation, new people with possibly different income levels. If we are making an index of proprietors' incomes, we want to assure that the index represents the cash flow generated by specific investments, and not new investments.

The standardization we seek in the indices used to settle contracts is essential to liquidity in these markets. Saying that markets create liquidity is almost the same as saying that they create markets for standardized assets. Illiquid markets tend to be markets where the individual assets are idiosyncratic, having quality characteristics that are unique to each asset sold, assets that are difficult to describe and measure. The market for any one narrowly defined commodity, where the definition includes the complete specification of type (quality and location), is therefore very shallow, even if a vast number of measures of prices or rents or incomes of all types together occur every day.

There is already a substantial literature on the issues of quality variations in the construction of consumer and producer price indices. But that literature is mostly concerned with producing indices of prices of newly produced commodities, not indices of the prices of claims on future incomes or services. The contributions made here to that literature concern the method of producing price or income indices for such individual claims, and extending the literature to cases where prices or incomes of individual subjects or items are observed infrequently (prices of houses are observed only when sold).

This chapter will first review some existing index number methods, and then extend these methods to deal with our problems. Chain index and hedonic index number methods will be reviewed, and ordinary repeated-measures indices (like the repeat sales indices) are shown to be in a sense a special case of these, and to have strong parallels to some existing indices used to settle contracts. The hedonic repeated-measures index is introduced to allow for control of changing price of quality variables, while retaining the repeated-measures design.

Through much of the discussion in this and the next two chapters, I will refer to the price index construction methods as applied to the problem of estimating real estate price indices, so that each item is an individual property and each measure is a price at a sale of that property, even though the intended applications are wider. Real estate is a good example of the applications of the index construction methods; the observations on price come only when a sale is made, which may be very infrequent, so the difficulties facing developers of index numbers may be particularly severe.

Analogies to other indices used to settle contracts

It is helpful first to look at existing indices that are used to settle contracts. This will be done, not to borrow a theoretical framework from elsewhere, but to illustrate what has worked for the purpose of contract settlement, and to provide some parallels to what will be developed below. Both stock price indices and consumer price indices are widely used to settle contracts. Stock price indices are used to settle futures, options, swaps, and other financial contracts. The consumer price index is used in cost-of-living clauses in many contracts, such as labor contracts.

Those who construct price indices for corporate stocks might also be considered, in a sense, to face a problem of assuring that their index applies to a standard item. The qualities of individual stocks are very hard to define. The individual stocks are not easily summarized—an appraiser who was asked to put a price on an individual stock would have a difficult time doing so without relying on market price of the particular stock in question. Fortunately, our stock markets are set up so that all stocks traded in a given firm are inherently claims on the same cash flow, and so we can, for major, publicly traded stocks, observe, for each time period, a price of a share for each firm. This allows us to create a price index for major stocks based entirely on regular repeat sales of shares of individual firms. One might think of the stock price indices as indices based on no information about qualities at all, but instead on observed repeat sales of each property (considering different shares of the same corporations as repeat sales of the same property), and construct an index based on changes in prices of individual properties.

The conceptual basis for value-weighted arithmetic (VWA) stock price indices, such as the Standard and Poor's Composite Stock Price Index,[1] is that the index should replicate the value of a portfolio that invests in all stocks in proportion to their values outstanding, a portfolio whose composition is rebalanced (shares are bought and sold) each period to keep it in correspondence with the amounts outstanding. The changes in value of such a portfolio are determined entirely by changes in prices of constituent stocks. Such an index is constructed, therefore, entirely from repeat sales data. Such a portfolio represents the market to the extent that its price changes are observable, and therefore is appealing as the basis of settlement for a contract that is used to hedge market risk. No rebalancing of the portfolio used to construct the index is necessary in response to price changes, since the value in a portfolio whose constituent stocks are, in the preceding period, in proportion to values of those stocks outstanding will stay in proportion to these values when only prices change. But the numbers of shares held in the portfolio must be changed every period that the numbers of shares outstanding change, as when there are new issues of shares or firm repurchases of shares.

The value-weighted arithmetic price index I_{VWAt} on any given time period t is determined from the index of the preceding time period according to the chain index formula:

$$I_{VWAt} = (\sum_i Q_{it-1}P_{it} / \sum_i Q_{it-1}P_{it-1}) I_{VWAt-1} \tag{6.1}$$

where P_{it} is the price of a share of stock in company i at time t, Q_{it-1} is the number of shares outstanding of the ith stock at time $t-1$, and the summations are over all companies i. The index is set at an arbitrary value, which will be taken here to be 1.00 (though 100 is more common) at the base period, which will be assumed here to be at $t = 0$. The ith element in the summation in the denominator is the value of all shares in the ith firm at time $t-1$. Therefore, the ratio $\text{Index}_t/\text{Index}_{t-1}$ is a value-weighted (V_{it-1}-weighted, where $V_{it-1} = P_{it-1}Q_{it-1}$) arithmetic average (averaging over firms, indexed by i) of the price ratios (price at time t over price at time $t-1$).

The chain index was first defined by Irving Fisher (1911) in the context of indices purporting to measure the cost of living. (The chain index was later given further motivation by Divisia (1925) who proposed a continuous time price index analogous to (6.1) as well as an analogous quantity index, such that the product of the indices was always proportional to the total value sold.) It is called a chain index because the index at time t is linked to the index for the preceding period using data between time $t-1$ and time t. Chain indices may be contrasted with the more conventional fixed-base indices that use a single set of weights Q_i, unchanged from period to period.

The same formula (6.1) can be rewritten by letting i index the share, rather than the corporation (so that corporations are no longer indexed separately, although in the data all shares of a given corporation at a given time will have the same price):

$$I_{VWAt} = \frac{\sum_{i \in q_t} P_{it}}{\sum_{i \in q_t} P_{it-1}/\text{index}_{VWAt-1}} \tag{6.2}$$

where q_t is the set of all shares that were outstanding at time $t - 1$. When the index is written in this form, we see that the value-weighted arithmetic index at time t is based on simple sums of prices of all shares outstanding at time $t - 1$: it is the sum of their prices in time t divided by the sum of their prices in the base year, where the index at time $t - 1$ is used to deflate the $t - 1$ period price to the base year. Or, we could say (dividing both numerator and denominator of (6.2) by the number of shares outstanding at time $t - 1$) that the index at time t is the average price at time t of all shares outstanding at time $t - 1$ divided by their inferred average price in the base period.

The index represented by (6.1) and (6.2) is the value of a portfolio of stocks, a portfolio that is adjusted in each period so that the values held of each stock are in proportion to the values outstanding that period. As such, the index is physically replicable. Index arbitragers can create a portfolio whose value tracks the index. In contrast, the first stock index futures contract, the Value Line contract at the Kansas City Board of Trade, used a geometric index of stock prices, which could not be physically replicated. Because the geometric average of a set of non-negative numbers not all the same is always less than the arithmetic average, there is a downward bias in such indices relative to arithmetic indices. Indeed, should the price of any single stock fall in value to zero, then the entire index would fall to zero. The geometric index has since been replaced by an arithmetic index in the Kansas City Value Line contract.

In some respects, the US producer price index (as also the consumer price index) may be regarded as a chain index in its finest components, the 'cell indices', and a fixed weight index in its aggregation of components into a broader index. In estimating a cell index, formula (6.1) is sometimes used (see US Department of Labor, 1986, p. 113), where i denotes a particular kind or model number of commodity and Q_{it} is the weight given to kind or model i in the index at time t, a weight that is abruptly set to zero when the kind or model is dropped. The Q_{it} applied to individual items used to compute the cell indices are changed constantly as models and firms sampled are changed. In contrast, the weights used to aggregate the cell indices into more aggregate indices are not changed (except at very long intervals of time). The chain index method is used to deal with the quality change. When a model of television sets, say, is discontinued, index constructors may replace it in their sample with a new model of television set and adjust by the ratio of prices of the old and new models. However, in applying equation (6.1) they may use a method called 'link to show no change' (see US Department of Labor, 1986, p. 98). In (6.1) the price of any discontinued model (for which Q_{it} is zero) is assumed not to change between period $t-1$ and t, so that P_{it} in the numerator is replaced in this circumstance with P_{it-1}. The reason is that it may be difficult to observe a period in which both models are sold; and, perhaps, even if they do observe such a period, the old model may be remaindered and on sale, so that if they used equation (6.1) they would create a downward bias in the index, by neglecting changes in the perceived quality of the old model when it has been discontinued.

While producer and consumer price indices are useful analogies to consider when constructing indices of prices of claims on flows of incomes or services, the theory behind the indices is different in important ways. With producer (or consumer) price indices, the quantity Q_{it} is supposed to represent the quantity of the *i*th good produced (consumed) in this period. With consumer price indices, the quantities Q_{it} are then the arguments of the utility function that a representative consumer is assumed to maximize given a budget constraint defined in terms of prices and income. In contrast, with indices like the stock price indices the quantities Q_{it} are quantities of investment assets held, rather than quantities consumed. For an example of a conclusion from the consumer price index literature that is not applicable to the present context, consider the statement that a problem with a chain price index is that the index at two non-adjacent time periods need not be the same even if all Q_{it} and P_{it} are the same in both periods (if the Q_{it} were different in an intervening period). That is not a problem in the present context: an investor who held all stocks in the proportion outstanding would not expect to have the same portfolio value in the two periods, because of the effects of the changed portfolio proportions between the two periods.

The value-weighted arithmetic stock price index may be contrasted to an index that is meant to correspond to the price of some 'representative' stocks. The Dow Jones Industrial Average is based on the average price of 30 stocks chosen to be representative of industrial stocks. The differences between the Dow Jones Industrial Average and a value-weighted arithmetic index like the Standard and Poor's Composite Index are due, of course to the different weighting given the stocks: weights do not correspond to value as in the formula above, and only certain stocks are weighted. The Dow Jones Industrial Average might be a better basis for cash settlement for a hedger whose portfolio consists of equal numbers of shares in each Dow stock, but cannot be expected to be as good for someone who has a more typical portfolio. Even an investor who has invested only in the Dow list of stocks might prefer cash settlement based on an arithmetic value-weighted stock price index representing the whole market, since such an index gives more weight to the stocks for which there are more shares outstanding. Through time, if the composition of the Dow portfolio is not changed, the index is likely to become less and less representative of the market. If the Dow stocks were initially chosen as the 30 most important stocks, then, through time, because of the various successes of different firms, they will tend not to continue to be the 30 most important stocks. The constructors of the Dow Jones Industrial Average deal with this problem judgmentally, by occasionally adding or dropping stocks. In contrast, such adjustments are in effect made continually by the chain index.

This problem of gradual outdating of any given portfolio of stocks is somewhat analogous to the problem that is faced by constructors of consumer price indices. A consumer price index, such as the ones published by the Bureau of Labor Statistics in the United States, is supposed to represent the price of a basket of goods through time that is representative of the expenditures of consumers. The problem is that the market basket consumed changes through time. Suppose, for example, that long ago, when caviar was abundant, it was widely consumed. Later, when it became scarce, much less of it was consumed, and its price became astronomical. A consumer price index that priced the old market basket may become unrepresentative of the prices people pay, since caviar is virtually no longer consumed. That problem is dealt with by, at long intervals of time, rebasing the index, and updating the market basket. It would be better, were it not costly to re-estimate the amounts consumed of each commodity every period, to use every month a chain index formula like (6.1) for the consumer price index.

Extending chain indices to infrequently traded assets

Chain indices like those discussed above are not directly applicable to prices of illiquid assets, such as houses, because the assets are not sold every period, and may be sold only very infrequently. Nor are the index methods capable, without some modification, of creating income or wage indices from data on individuals or households that are collected on a rotation basis, sampling at intervals from the same households, as for example with the Current Population Survey. Since data are costly to collect, data on incomes are efficiently collected on such a rotation basis. The indices from the preceding section will nonetheless serve as an inspiration for some of the indices that we can construct. In what follows, we shall seek the best analogues to these stock price indices or consumer price indices for application to such data.

The formula (6.2) is more natural than (6.1) to apply to non-standardized subjects like houses, in that there is no natural analogue to the firm in the house example. If there were a finite number of different kinds of houses, and if each house of a certain kind had the same price at time t, then we could regard each kind of house as the analogue of a firm, and use (6.1), where Q_{it-1} is the number of houses existing at time $t-1$ of kind i. But prices of individual houses are not well determined by their type. Nor are the prices accurately determined by any formula in terms of observable characteristics variables. Prices of individual properties are not entirely predictable on the basis of any information we have; any categorization we have is imperfect, not defining quality fully, and we always run the risk that some unobserved quality variable is changing through time.

We cannot apply (6.2) directly, since most of the prices in the formula are unobserved. In the real estate application we can observe a price only in the time periods when a property is sold. A natural analogue to (6.2) that can be applied to existing data is the chain index I_{Ct}, and which will reappear in the more formal analysis below:

$$I_{Ct} = \frac{\sum_{i \in q_t} P_{it}}{\sum_{i \in q_t} P_{it_{it}} / I_{Ct_{it}}} \tag{6.3}$$

where q_t is the set of all subjects for which there are measures (observed prices at time of sale) for the second or later time at time t, and t_{it} is the time of prior measure (sale of property i). This formula is analogous to (6.2) and reduces to it if t_{it} equals $t - 1$ for all i. This formula, as with (6.2), if applied consistently since the beginning of the index, represents in the real estate application the value of a portfolio of properties. Suppose that at time 0, the base period, the index is set to 1.00. Then at time 1, the only sales in q_t are properties that were sold at both time t and $t - 1$; the index then is the total value of those properties at time 1 divided by the total value of those properties at time 0; in period 1 the index is clearly the value of a portfolio consisting of one dollar invested in all these properties. At time 2, there are properties that sold at time 0 and time 1, time 0 and time 2, and time 1 and time 2. The denominator of this index for time period 2, then, is the time 0 value of all these properties. Now, the property that sold at time 1 and time 2 is divided in this formula by the index at time 1; its base-year value is the value of an investment in all properties at time 0 that were sold again at time 1, reinvesting the proceeds at time 1 into the property that sold at times 1 and 2. At times 3, 4, and beyond, by the same reasoning, we see that the index represents the value of a portfolio of properties. It is not, however, a portfolio that can be physically replicated in real time, since one cannot buy shares in illiquid individual properties, and in any event cannot know in advance when a property will later sell.

The portfolio represented by (6.3) is value-weighted, in the sense that it gives more weight, in determining percentage price index changes, to the percentage changes in the more valuable properties. This is desirable if we want the index to represent the market, since the more valuable properties represent a bigger component of the market, and may show a different price path through time from that followed by the less valuable properties.

The portfolio in (6.3), while analogous to (6.2), represents a rather older portfolio of properties. The newest properties, such as houses built recently, are less likely to have been sold twice, and so such properties are excluded from any input into the index. Indeed, there is no way to use these new properties sold only once in an index analogous to (6.2), since there is only one price observed for each of these. To make use of these properties in an index, one would have to abandon the notion that changes in the index will be determined only by changes in prices of individual properties. This would seem to be a particularly dangerous thing to do, since new houses, having been built recently, are, most of them, likely to embody in common some quality changes that are unmeasured. These unmeasured quality changes are likely to be systematic, i.e. applying to all of these new houses, and hence might cause important biases in the index.[2]

In using (6.3), it should be borne in mind that there is a risk that the sales of properties for the second time on a certain date need not be representative of the different kinds of properties that exist. The number of large houses sold, for example, may be lower than usual in a certain time period, even if the number of such houses that exist is not changed. This risk can be dealt with if there are data about the qualities of houses sold, as will be shown below.

If we do have some information about qualities of properties sold, such as their classification into categories of properties, and if we also know the relation between qualities and numbers of properties existing (data such as the number of existing properties in each category), then we can begin to correct a price index for possible variation in the representativeness of the houses sold. We could, for example, estimate price indices for each of a number of different kinds of houses, and then, using the data on the number of houses existing at time t in category i as Q_{it}, produce a chain index using (6.1), where the price P_{it} of kind i of property at time t is some price index using data on properties of this kind. In that case, we might still want to use an index like (6.3) for each of the price indices that would be used as inputs to (6.1); since prices are not identical for all properties of a given kind, some properties showing much more value than others, we will still want to use a value-weighted formula for each property type.

In order to formalize these notions of index numbers for subjects whose prices or rents are observed infrequently or at intervals, it is important to develop a statistical model of these prices, so that a sampling theory can be developed. It is helpful to do this first in a rather more general context than is suggested by the simple index formula (6.3), so that we can develop indices that correct for time variation in the qualities of the subjects that are observed, as for example a change in the mix of properties sold. To do that, the hedonic regression model will be developed first.

Basics of hedonic indices

The various methods of dealing with quality changes that are practically implemented today may all be described as variations on the hedonic regression method originally proposed by Court (1939). Court's methods were generally neglected until a revival of interest was spurred by Griliches (1961); still, they are not widely used by those who produce economic indices for regular publication; see Triplett (1990). I shall here discuss the fundamentals of their construction, the problems they have which have inhibited their use. This discussion will lead us to a hedonic repeated-measures index, which can help deal with these problems.

The regression-per-period hedonic (log) index is produced by regressing, for each period, log price (or log income or log of another measure) on a constant and a number of variables, called hedonic or quality variables, that characterize the item to which the observation applies. In our real estate price example, each observation represents a sale of a property. The dependent variable is the log price of this property at time of the sale, and the hedonic or quality variables correspond to characteristics of this property at the time of the sale. Let us use Y_t to denote the dependent variable, a column vector having N_t elements, where N_t is the number of sales observed at time t. The element p_{it} of Y_t is the (natural) log price of the sale of property i at time t. In the real estate example, we might want to subtract from the price at time t of the property i the present value at time t of the past expense of the investments made in maintaining the property from some reference date in the past until time t, and add to it the present value of the rents received on the property, so that we have the total value of an investment in the property; but normally, in real estate applications, such data are unavailable.[3] Let us use Z_t to denote the matrix of independent variables, an $N_t \times K$ matrix whose *i*th row consists of a vector of quality variables for that date.[4] In the context of real estate prices, if $N_t = 3$ and Z_{it} consists of a constant and s_{it}, the (natural) log number of square feet of floor space for that property at time *t*, then an example of these matrices is:

$$Y_t = \begin{bmatrix} p_{1t} \\ p_{2t} \\ p_{3t} \end{bmatrix}, \quad Z_t = \begin{bmatrix} 1 & s_{1t} \\ 1 & s_{2t} \\ 1 & s_{3t} \end{bmatrix}. \tag{6.4}$$

The regression model is $Y_t = Z_t\gamma_t + \varepsilon_t$, where ε_t is the vector of regression error terms, assumed to have zero mean and to be independent of s_{it} for all *i*. Index numbers will then be generated using the estimated coefficient $\hat{\gamma}_t = (Z_t'Z_t)^{-1}Z_t'Y_t$. If there is heteroskedasticity, or other deviation from the spherical normal assumption for the residuals, then we may postulate a variance matrix Ω_t for the error vector at time *t*, and use instead a generalized least-squares estimate $\hat{\gamma}_t = (Z_t'\Omega_t^{-1}Z_t)^{-1}Z_t'\Omega_t^{-1}Y_t$.

Let us combine these regressions for *T* time periods, $t = 0, ..., T-1$, into one giant regression. To do this, we create an *N*-element *Y* vector (where $N = \Sigma n_t$) by stacking the Y_t, $t = 0, ..., T-1$, and we create a block-diagonal *Z* matrix of dimension $N \times TK$, with *T* blocks Z_t, $t = 0, ..., T-1$. Moreover, we create an $N \times N$ variance matrix Ω; this might be proportional to the identity matrix, which would make generalized least squares reduce to ordinary least squares, or it might be a block-diagonal matrix with the blocks Ω_t, $t = 0, ..., T-1$, or it might have some other form. Then a single regression will compute price indices for all times for which we will have the index. Supposing, to economize on space, that there are only three time periods, times 0, 1, and 2; then the giant regression is:

$$Y = \begin{bmatrix} Y_0 \\ Y_1 \\ Y_2 \end{bmatrix}, \quad Z = \begin{bmatrix} Z_0 & 0 & 0 \\ 0 & Z_1 & 0 \\ 0 & 0 & Z_2 \end{bmatrix}. \tag{6.5}$$

Since the *Z* matrix is block-diagonal, then if the Ω matrix is also block diagonal, the estimated general least-squares coefficient vector $\hat{\gamma} = (Z'\Omega^{-1}Z)^{-1}Z'\Omega^{-1}Y$ is just the stacked per-period regression. Since the index is intended to be produced in real time, a value at a time, we would keep expanding the Y and Z matrices by appending the latest Y_t to Y and the latest Z_t to the bottom right corner of a matrix whose rows and columns have been augmented from Z with a row and column of zeros. This entails no revisions in past values of the index computed before.

A Laspeyres price index, using base-period quality variables, can be constructed from the estimated coefficient vector $\hat{\gamma}$:

$$I_{\text{Laspeyres}} = \begin{bmatrix} \bar{Z}_0 & 0 & 0 \\ 0 & \bar{Z}_0 & 0 \\ 0 & 0 & \bar{Z}_0 \end{bmatrix} \hat{\gamma} \tag{6.6}$$

where $\bar{Z}_0$ is a $1 \times K$ vector of quality variables for the base period, period 0. Now, of course, for many applications the average quality of items sold in the base period may not be representative of the items we wish to measure. For example, in constructing a house price index we may prefer, not the average quality of houses sold in the base period, but the average quality of houses outstanding in time t. Houses sold may be systematically unrepresentative of all houses; house sales probably over-represent newly constructed houses, for example, as noted above. One may also, having produced the single regression estimate $\hat{\gamma}$, wish to derive a number of indices from it, representing different kinds of assets, and there might be a futures market for each such index. For example, we might, if computing real estate price indices, want separate indices for small and large houses, and a vector $\bar{Z}$ might be designed for each of these.

There is also the issue of constructing a chain index, as discussed above. The quality vector of outstanding houses, for example, may change through time, and we want the index to remain relevant through time to the average house. In this log-price index, the regression per period chain index $I_{rppchain}$ looks a little different from above, in that it is additive rather than multiplicative:

$$I_{rppchaint} = I_{rppchaint-1} + \bar{Z}_{t-1}(\gamma_t - \gamma_{t-1}) \,. \tag{6.7}$$

The regression-per-period method is not the only way of constructing hedonic price indexes, and, indeed, other methods, representing restricted forms of the regression-per-period method, may be more common. There may be substantial multicollinearity among quality variables of assets sold at a particular time, so that the fitted value for an asset of standard quality in a given period may, if the standard asset was not sold much in that period, behave very erratically. Researchers may reject the regression-per-period method from bad experience, having observed sometimes bizarre (even negative) values of such price indices.

Problems in measuring quality

In constructing such hedonic indices, one is inevitably struck by the arbitrary or judgmental decisions one must inevitably make. Not only is there the decision of which quality variables to include, but there are also decisions to make about allowing nonlinear effects of each and interaction effects (represented, say, by variables equal to products of characteristic variables) between them. One researcher might overlook a nonlinear term or interaction effect that another chooses to include, and including this variable could certainly reverse the trend of some indices. A researcher who wants to stir up controversy by producing very different indices could always add some variables that are highly collinear with others, thereby widening standard errors of estimated coefficients. This tactic can produce estimated indices that are likely to show little resemblance to other indices. With so many choices to make, an unscrupulous person could tell a research assistant to explore the many different possible combinations of hedonic variables until the index looked as wanted. There is a fundamental problem of objectivity to such indices. Of course, such problems are not necessarily unmanageable; there is a problem of objectivity in coffee-tasting too, and futures markets do exist for coffee.

The other problem with hedonic methods is simple lack of data on quality. We rarely can observe all the hedonic variables that are relevant to valuation. There are two aspects to this information problem: sample size and unobserved characteristics.

For example, we cannot expect to find large samples for which a large number of characteristics of houses, such as square feet or architectural style, is readily available. It is costly to find such data. Finding such data may incur certain lags in reporting, but most importantly, such data are just nowhere available for all houses. Multiple-listing services make some such data available, but only for a small list of characteristics, and only for the subsample of houses that are listed. Any attempt to get a representative sample of house sale prices with a large list of carefully measured characteristics will of necessity yield a small number of observations. A small number of observations means high standard errors on the index, as well as a bigger potential for selection bias problems.

The problem of unobserved characteristics is that we can never be sure that another quality variable will not be discovered later that will, if added to the regression, cause dramatic revisions in the index. We can never be sure that some of the coefficient values are not due to omitted variables. Consider, for example, the Constant Quality Index of real estate prices produced by the US Department of Commerce. In this regression-per-period index, in recent years, the coefficient on the air conditioning dummy in indices for the west of the United States has been negative. It seems nonsensical that homes with air conditioners would be worth less than homes without. One hypothesis for the negative coefficient on the air-conditioning dummy is that homes without air conditioning tend to be located in regions with cooler weather, such as along the shore, and so the air conditioning dummy is negatively correlated with another (at present unmeasured) quality variable. This negative correlation might also change through time, as air conditioners become more common, and this change would create a systematic bias in the growth in an estimated hedonic price index for a standard house with air conditioning.

For another example of the kinds of problems we may find with unobserved quality variables, consider the hedonic price indices for automobiles that used the weight of the vehicle as a measure of quality. When in the last decade fuel economy became an important selling point for cars, manufacturers downsized their new models. Hedonic regression-per-period price indices for cars of standard weight would then show spurious price inflation over the time period that new cars were getting smaller. Clearly, weight is not a measure of quality; it is merely correlated with quality. There is actually no readily obtained measure of the quality of automobiles other than their price. (Hence, factor-analytic models will be proposed in the next chapter that estimate quality from price.)

The difficulties that we face in specifying hedonic variables should not be reason to avoid their use altogether, however. It is not even meaningful to say that one can avoid their use altogether, since when one estimates indices one must inevitably decide which subjects to include and which to exclude, a decision which in itself can be described as incorporating dummy hedonic variables into the analysis. Rather, the lesson we should learn from the difficulties that people have had with hedonic methods is that we should try to find index construction methods that are relatively robust to omission of hedonic variables; and this brings us to the repeated-measures indices.

Repeated measures and hedonics

The repeated-measures indices to be developed here are based on a regression model that differs from the giant regression model (6.5) only by the addition of dummy variables that proxy for the omitted hedonic variables. These additional variables may be called subject dummies, as in the literature on experimental design, where the experiment under consideration often involves human subjects, although they might better be called property dummies or asset dummies in some of our applications. Each subject has its own dummy variable; for example, there is one subject dummy for each individual property if we are estimating a real estate price index.

The term 'repeated measures' is commonly used in the literature on experimental design to refer to methods that rely on observing the changes in the effects on individual subjects; see for example Lee (1975). An experiment designed to test the effects of a drug may consist of giving the drug to an experimental group and giving a placebo to a control group, to verify whether there is any difference in the response. The experiment has a repeated-measures design if the same subjects are studied on both groups (at different times). By confining the analysis to observations of the same subjects in both groups, one rules out the possibility that the results are due to differences between the two groups. The repeated-measures design is especially important if it is impossible to control the membership of the two groups, as when some subjects drop out of the study in response to the drug. Such a repeated-measures design is at least as important in some of our index number calculations as in these medical experiments. When we are producing indices based on infrequent observations of real estate sales or the like, and are trying to learn about differences through time rather than between control and experimental groups, then we certainly have the problem that we cannot control which subjects are observed on each date.

Repeated measures of the same subjects have also seen use among economists who use panel data, see for example Hsiao (1986). The transformation to be developed here for our hedonic repeated-measures index resembles the kind of differencing employed in many panel data analyses, although here the differencing intervals are dictated by our observations, and the differencing intervals tend to be variable and overlapping.

In the hedonic repeated-measures model here, the dummy variable corresponding to the *i*th subject is zero unless the observation corresponds to that subject, in which case the value is 1.00. Of course, the sum of all these dummies is the vector of ones, and so, to avoid multicollinearity, one of the constant terms in the Z matrix can be dropped. Moreover, if any hedonic variable is constant through time for all subjects, then the subject dummies will be collinear with the columns of the Z_t matrices, $t = 0, 1, ..., T-1$, corresponding to this hedonic variable. The sum of the subject dummies multiplied by the square feet of floor space of the corresponding property equals the sum of all the columns of the Z that correspond to square feet measures. So, we must drop some columns, call the number of columns dropped h, and we choose here to drop columns of Z_0.

It is easiest to show the method of using subject dummies by a real estate example, where the dimensions of the matrices are much smaller than we would expect in applications, to keep the dimensions of the matrix small:

$$\begin{bmatrix} p_{10} \\ p_{20} \\ p_{30} \\ p_{40} \\ p_{11} \\ p_{21} \\ p_{51} \\ p_{32} \\ p_{42} \\ p_{52} \end{bmatrix} = \begin{bmatrix} 0 & 0 & 0 & 0 \\ 0 & 0 & 0 & 0 \\ 0 & 0 & 0 & 0 \\ 0 & 0 & 0 & 0 \\ 1 & s_{11} & 0 & 0 \\ 1 & s_{21} & 0 & 0 \\ 1 & s_{51} & 0 & 0 \\ 0 & 0 & 1 & s_{32} \\ 0 & 0 & 1 & s_{42} \\ 0 & 0 & 1 & s_{52} \end{bmatrix} \begin{bmatrix} \gamma_{11} \\ \gamma_{12} \\ \gamma_{21} \\ \gamma_{22} \end{bmatrix} + \begin{bmatrix} 1 & 0 & 0 & 0 & 0 \\ 0 & 1 & 0 & 0 & 0 \\ 0 & 0 & 1 & 0 & 0 \\ 0 & 0 & 0 & 1 & 0 \\ 1 & 0 & 0 & 0 & 0 \\ 0 & 1 & 0 & 0 & 0 \\ 0 & 0 & 0 & 0 & 1 \\ 0 & 0 & 1 & 0 & 0 \\ 0 & 0 & 0 & 1 & 0 \\ 0 & 0 & 0 & 0 & 1 \end{bmatrix} \begin{bmatrix} \delta_1 \\ \delta_2 \\ \delta_3 \\ \delta_4 \\ \delta_5 \end{bmatrix} + \begin{bmatrix} \varepsilon_{10} \\ \varepsilon_{20} \\ \varepsilon_{30} \\ \varepsilon_{40} \\ \varepsilon_{11} \\ \varepsilon_{21} \\ \varepsilon_{51} \\ \varepsilon_{32} \\ \varepsilon_{42} \\ \varepsilon_{52} \end{bmatrix} \quad (6.8)$$

In this example, $T = 3$ (two time periods after the base period 0), $K = 2$ (the Z matrices contain two columns, a column corresponding to the constant term and a column representing a hedonic variable), $N = 10$ (10 sales are observed), $k = 5$ (there are 5 properties), and there are 2 sales observed for each of the properties, so that there are $n = 5$ repeated-measures pairs. As before, p_{it} represents the (natural) log price of the ith property in time t, and, as before, s_{it} denotes the (natural) log number of square feet of floor space of property i at time t, our single hedonic variable. Here, the error term ε_{it} for property i at time t is due to the variation in market price of the individual property and noise in the house sale process; such things as randomness in the efforts made to sell a property, the random arrival of interested buyers, or the fluctuations in neighborhood conditions that are unrelated to fluctuations in conditions in the entire geographical area for which the index is constructed. The error terms ε_{it} are arranged here into a N-element vector ε which is assumed to be distributed independently of the dummy variables, and is assumed to have mean 0 and variance matrix Ω. Here, δ_i is the coefficient of ith subject dummy, δ_i is an indicator of the 'quality' of the ith property that is not captured by the square foot variable.

Note that to use this framework we do not necessarily need to have the data that would give the actual square feet of floor space for each house. We could replace for s_{it} in (6.8) with the average number of square feet of houses in the region in which house i is situated. Doing this may be very useful; for example, in the United States census data are available giving the average number of rooms in houses by census tract, zip code, or county, although only at decades. Using such data would not work if we have only one region for which the aggregate improvements data are available and if we are computing an index for this same region, since there would be multicollinearity between the columns of z corresponding to the s_{it} data and the columns of the z that contain only ones and zeros. However, if we have data on square feet in at least two sub-regions of the region for which the price index is to be computed, then not all properties will be represented in z as having the same square feet, and there will generally not be strict multicollinearity. Of course, one might say that there is now a measurement error in the square feet of floor space variable, in that each property is assigned the average for its region rather than its own square feet of floor space, but this measurement error will not destroy consistency in regression estimates because the error is uncorrelated with the measured variable rather than with the true number of square feet. Alternatively, one might say that replacing the square feet with its region mean is analogous to estimating (6.8) with actual square feet for each property but with a two-stage least-squares method with instruments equal to region dummies. In two-stage least squares, the independent variables are first regressed on their instruments and then the fitted values are used in a second stage regression; here, the instruments would be region dummies and the fitted values the region mean square feet of floor space, so that the second stage of the two-stage least-squares method would be the same as a regression y on z where z was constructed using the regional average square feet rather than actual square feet of each house. How well this method works, of course, depends on how much interregional variation there is in the number of square feet of floor space; if there is not much variation, then there will be near multicollinearity among the columns of z, and the standard errors of the estimated coefficients will be high.

The regression model exemplified by (6.8) resembles the usual fixed-effects analysis of covariance model. In the analysis-of-covariance literature, the subject dummies would be called experimental factors, and the hedonic variables the concomitant factors. Of course, the motive of analysis-of-covariance is usually centered on discovering the coefficients of the experimental factors while here we regard the coefficients of the subject dummies as nuisance parameters, our interest centering instead on the γ coefficients.

In this example, the number of columns h of Z_0 that are dropped to avoid multicollinearity equals 2, reflecting our assumption here that none of the properties shown had a change in the number of square feet of floor space in the sample period, so that there would be multicollinearity if either column of Z_0 were included. The matrix Z_0 has thus been dropped completely from the giant matrix Z; had there been properties that had had additions put on, so that their square feet of floor space had been changed, then we would have retained the second column of Z_0 in Z and retained a coefficient for this column as an element of the vector of coefficients γ.[5]

The reader may well wonder, at this point, why I am retaining the square feet of floor space variable at all when adding these dummies, since, so long as there is no change in the square feet of individual properties, the dummies span the information in the combined (summed) square feet variables. But they do not span the set of vectors (including columns 2 and 4 of the matrix that multiplies γ in (6.8)), one for each time period, which are needed if the regression model is to allow time variation in the response of price to the square feet variable. The reader may also wonder why I adopt the convention of dropping, when columns must be dropped because of multicollinearity, the columns of Z_0. It is convenient to drop columns corresponding to the base period since the index (which is in logs) will be set to zero in the base period. Dropping the column of Z_0 corresponding to a hedonic variable in this setting means that the coefficients corresponding to the remaining columns of Z representing this hedonic variable will show the change through time from the base period of the impact of this hedonic variable on price; the subject dummies show the impact of the unchanging quality of the property on price.

Simplifying regression forms

Let us rewrite this model in the form $Y = Z_A B + \varepsilon$ where Z_A is the matrix formed by combining the two independent variable matrices in the model, and B is a $TK - h + m$ element vector of coefficients where m is the number of subjects. The dimension of

Z_A is $N \times TK - h + m$. A generalized least-squares estimate of B is $\hat{B} \equiv (Z_A'\Omega^{-1}Z_A)^{-1}Z_A'\Omega^{-1}Y$; the ordinary least-squares case described above is of course given by the same formula where Ω is proportional to the identity matrix.

Now, the addition of the subject dummies as columns to produce the Z_A matrix above will generally break the block diagonality of the matrix $Z_A'\Omega^{-1}Z_A$, and will cause the method to generate later revisions of the index, to be produced as the matrices are augmented with the new data that arrive as time passes, even if the data used to generate past values of the index were complete, i.e. the new data consist only of properties sold for the second time in the new time period. These revisions arise even in the case of ordinary least squares or where the variance matrix Ω is block-diagonal. The appearance of such revisions would seem to be a disadvantage when compared to the regression-per-period methods, that, when ordinary least squares is used, do not produce revisions. Revisions of the data may be troublesome, as will be discussed in Chapter 8.

The $N \times (TK - h + m)$ Z_A matrix may have a very large number of columns; there is a column for every subject. Thus the $Z_A'\Omega^{-1}Z_A$ matrix can have large dimensions, and inversion of this matrix may be troublesome. But we want the coefficients of the Z_t variables rather than the coefficients of the dummy variables.

We can rearrange the setup of the regression to eliminate the need to invert large matrices. Since we do not need the values of the coefficients of the dummy variables to construct index numbers, we can eliminate them from our estimates. To do this, we construct the nonsingular $N \times N$ matrix $S = (S_1', S_2')'$. In the example, S_1 is 5×10, S_2 is 5×10, and the matrix S is:

$$S = \begin{bmatrix} -1 & 0 & 0 & 0 & 1 & 0 & 0 & 0 & 0 & 0 \\ 0 & -1 & 0 & 0 & 0 & 1 & 0 & 0 & 0 & 0 \\ 0 & 0 & -1 & 0 & 0 & 0 & 0 & 1 & 0 & 0 \\ 0 & 0 & 0 & -1 & 0 & 0 & 0 & 0 & 1 & 0 \\ 0 & 0 & 0 & 0 & 0 & 0 & -1 & 0 & 0 & 1 \\ 1 & 0 & 0 & 0 & 0 & 0 & 0 & 0 & 0 & 0 \\ 0 & 1 & 0 & 0 & 0 & 0 & 0 & 0 & 0 & 0 \\ 0 & 0 & 1 & 0 & 0 & 0 & 0 & 0 & 0 & 0 \\ 0 & 0 & 0 & 1 & 0 & 0 & 0 & 0 & 0 & 0 \\ 0 & 0 & 0 & 0 & 0 & 0 & 1 & 0 & 0 & 0 \end{bmatrix}. \tag{6.9}$$

The matrix S_1 is so constructed that the ith element of $y = S_1Y$ is the difference of the ith pair of consecutive observations of the dependent variable for the same subject. If subject j appears three times, in times t_1, t_2, and t_3, then we will have two elements of $y = S_1Y$ for subject j, an element $p_{jt3} - p_{jt2}$ and an element $p_{jt2} - p_{jt1}$. The matrix S_1 will then be of dimension $n \times N$ where n is the number of pairs of consecutive observations of subjects that are constructable from the N observations. The $m \times N$ matrix S_2 is so constructed that $\tilde{Y}_2 \equiv S_2Y$ is the vector of all first observations of subjects. Let us call $\tilde{Y} = SY$, $\tilde{Z} = SZ_A$, and $\tilde{\Omega} = S\Omega S'$. Let us denote the upper left $n \times (TK - h)$ corner of $\tilde{Z}$ as z, the upper $(TK - h)$-element vector of B as γ, and the upper left $n \times n$ block of $\tilde{\Omega}$ as ω. Our hedonic repeated-measures index will be derived by regressing the first component y of $\tilde{Y}$, the component that equals S_1Y, onto the first component z of $\tilde{Z}$, the component that equals S_1Z_A, using a generalized least-squares regression where the variance matrix of the regression error terms is ω, the upper left corner of $\tilde{\Omega}$.

It can then be shown that the generalized least-squares estimate $(z'\omega^{-1}z)^{-1}z'\omega^{-1}y$ is the first part $\hat{\gamma}$ of $\hat{B} = (Z_A'\Omega^{-1}Z_A')^{-1}Z_A'\Omega^{-1}Y$. $\tilde{Z} \equiv SZ_A$ is block triangular. Its upper right $n \times m$ block consists only of zeros. Its lower right $m \times m$ block is the identity matrix. Letting $\Sigma \equiv \tilde{\Omega}^{-1}$ (which equals $S'^{-1}\Omega^{-1}S^{-1}$), the generalized least-squares estimator $\hat{B} = (Z_A'\Omega^{-1}Z_A)^{-1}Z_A'\Omega^{-1}Y$ can be written $(\tilde{Z}'\Sigma\tilde{Z})^{-1}\tilde{Z}'\Sigma\tilde{Y}$. If we now partition Σ into four square parts, with upper left $n \times n$ part called Σ_{11}, etc., then from the last m rows of $\tilde{Z}'\Sigma\tilde{Z}\hat{B} = \tilde{Z}'\Sigma\tilde{Y}$ (the last m normal equations for $\hat{B}$), one derives that $\hat{B}_2 - \tilde{Y}_2 + Z_{21}\hat{\gamma} = \Sigma_{22}^{-1}\Sigma_{21}(y - z\hat{\beta})$. If we substitute this into the first $TK - h$ normal equations we discover that $z'(\Sigma_{11} - \Sigma_{12}\Sigma_{22}^{-1}\Sigma_{21})z\hat{\gamma} = z'(\Sigma_{11} - \Sigma_{12}\Sigma_{22}^{-1}\Sigma_{21})y$, and so $\hat{\gamma} = (z'\omega^{-1}z)^{-1}z'\omega^{-1}y$ as was to be shown.

Ordinary repeated-measures indices

Before interpreting our hedonic repeated-measures index based on the regression coefficient $\hat{\gamma} = (z'\omega^{-1}z)^{-1}z'\omega^{-1}y$, let us first look at a degenerate special case of the regression in which the Z_t matrices do not include any hedonic variables, each matrix consists only of a single column of ones. Thus, we here delete the s_{it} variables in example (6.8). Lacking any hedonic variables, the method produces what we might call ordinary repeated-measures indices, or what have been called repeat sales indices.

The model here is just the usual two-way analysis of variance model; however, in the index number application we normally expect no more than one observation per cell (a cell denoting subject and date), and expect most cells to have no observations. It is possible that, when estimating an annual real estate price index, we might find occasionally that a property sold more than once in a year, but this would be unusual.

Using this expression as above, we can write a simpler expression for the regression model that is estimated to produce ordinary repeated-measures indices. The ordinary repeated-measures (repeat sales) regression model is $y = z\gamma + \varepsilon$ where the matrices z and y are:

$$z = \begin{bmatrix} 1 & 0 \\ 1 & 0 \\ 0 & 1 \\ 0 & 1 \\ -1 & 1 \end{bmatrix}, \quad y = \begin{bmatrix} p_{11} - p_{10} \\ p_{21} - p_{20} \\ p_{32} - p_{30} \\ p_{42} - p_{40} \\ p_{52} - p_{51} \end{bmatrix}. \tag{6.10}$$

The same regression model can be written in an alternative form, a difference form, such that estimated coefficients represent period-to-period changes in the log price level, rather than the price level itself. We replace the z matrix with a matrix whose ijth element is 0 unless, for the ith repeat sales pair, j is greater than the first sale date and not greater than the second sale date, in which case the element is 1. In this form, the ordinary repeated-measures regression is $y = z_d\gamma_d + \varepsilon$ where y is as above and the matrix z_d and vector γ_d are given by:

$$z_d = \begin{bmatrix} 1 & 0 \\ 1 & 0 \\ 1 & 1 \\ 1 & 1 \\ 0 & 1 \end{bmatrix}, \quad \gamma_d = \begin{bmatrix} \gamma_1 - \gamma_0 \\ \gamma_2 - \gamma_1 \end{bmatrix}. \tag{6.11}$$

The two forms of the model are, of course, identical, and yield identical estimates. This model was used by Bailey, *et al.* (1963),

Case and Shiller (1987; 1989; 1990), Webb (1988), Goetzmann (1992), Abraham and Schauman (1991), and others; see also Haurin *et al.* (1991).

The ordinary repeated-measures regression model can be interpreted as asserting that the change in price of each subject is the change in the aggregate price between sale dates plus an error term. Note that the conformation of the dummy variables in the z matrix is such that the ith element of $z\gamma$ is the change of the index between the two sale dates, and the corresponding element of y is the change in log price between the two dates. Since there are no hedonic variables, the fitted value of the regression for time t must be just $\hat{\gamma}_t$ and so our price index for time t is just $\hat{\gamma}_t$. No value of $\hat{\gamma}_0$ was estimated in the regression, but, since the base period for the log index is zero, normalizing the level index to 1.00 in the base period is the same as setting the log index for period zero at zero.

There is no constant term in the regression, because the model gives no reason to have an expected change in price that is independent of the dates of sale. Goetzmann and Spiegel (1992) find that, with the data set on four US cities used by Case and Shiller (1987), a constant term is statistically significant and positive in such a regression (using a heteroskedasticity correction for the error term). They interpret the constant term as reflecting improvements that the homeowner makes after purchasing a house, improvements that are thus made in response to a sale. If their interpretation is right, then one might include the constant term in order to take out this effect from the estimated index, producing an index that corrects for those improvements that occur at time of sale (though, of course, not for all improvements). On the other hand, the constant term might also have other interpretations, which would suggest different ways of constructing indices. It is possible that the estimated constant term is a consequence of heterogeneity of markets in the sample used to construct the index. When an index is constructed for a geographical area, the regression model assumes that all houses in that area follow the same price path, except for random noise. In fact, there are likely to be submarkets within the area that peak and trough at different times. Indeed, we sometimes estimate indices using the repeated-measures indices both for separate counties within a metropolitan area and then for the entire metropolitan area; doing this involves some inconsistency of modelling assumptions, since the metropolitan-area model does not recognize difference across counties. Heterogeneity might create what could be thought of as an errors in variables problem that may bias the slope coefficients in the regression. All the Goetzmann–Spiegel data sets were in periods when prices were increasing through time; since the mean price increase over their entire sample was positive, a downward bias in the slope coefficients should translate into an upward bias in the constant term.[6] If this is the right interpretation of their result, then we would not want to take out the effect of the constant term in computing an index; doing so would result in a sort of bias in the trend.[7]

Had our real estate price variables included the present value of rents received on the property and had the present value of maintenance and other investments made been subtracted, then the dependent variable in the regression would be the total return (continuously compounded, since we are showing logs of prices) between sales. Then, the index computed from this regression would be a measure of total value of an investment in real estate that reinvests all earnings in more real estate.[8]

The normal equations for $z'z\hat{\gamma} = z'y$, when ordinary least squares is used to produce the index $\hat{\gamma}$ are:

$$\hat{\gamma}_1 = \frac{p_{11} + p_{21} + p_{51}}{3} - \frac{p_{10} + p_{20} + (p_{52} - \hat{\gamma}_2)}{3} \tag{6.12}$$

$$\hat{\gamma}_2 = \frac{p_{32} + p_{42} + p_{52}}{3} - \frac{p_{30} + p_{40} + (p_{51} - \hat{\gamma}_1)}{3} \tag{6.13}$$

Note that, with these equations, the index $\hat{\gamma}_t$ is the average log price at time t of all properties sold at time t for which there is a pair of sales, minus the average of their other log sale price converted to base year by the index. In the above example, four of the five properties were sold in period 0; these did not have to be corrected to base-year price. Property 5, on the other hand, was not sold in period 0, and so for this property there is a correction, made by subtracting the index $\hat{\gamma}_1$, to convert p_{51} to base-year prices. If we rearranged the terms so that there is a term for each property, then we could see that $\hat{\gamma}_t$ is just average of all informative ways of inferring the price change between time 0 and time t, taking as given other index values. There are three ways of inferring γ_1 (using price changes of properties 1, 2, and 5) and three ways of inferring γ_2 (using price changes of properties 3, 4, and 5). Properties 3 and 4 are of no help in inferring γ_1 given γ_2, since they were not sold in period 1; taking γ_2 as given would enable us to convert the second price to a period 0 price, but that is still of no help in inferring γ_1.

Note that there is some analogy between the second normal equation (6.13) and the chain index (6.3) described above, and thus there is an analogy between the index and the value of a portfolio of housing; the analogy will become stronger when we move to arithmetic variants of the index below. This simple analogy is important when we consider the robustness to departures from model assumptions, such as the heterogeneity discussed above. But the (antilog of the) above index is geometric, in that it is based on logs of prices.

The same normal equations can also be written in the form $z_d'z_d\hat{\gamma} = z_d'y$:

$$\begin{aligned}\hat{\gamma}_1 - \hat{\gamma}_0 = ((p_{11} - p_{10}) &+ (p_{21} - p_{20}) \\ &+ (p_{32} - p_{30} - (\hat{\gamma}_2 - \hat{\gamma}_1)) \\ &+ (p_{42} - p_{40} - (\hat{\gamma}_2 - \hat{\gamma}_1)))/4\end{aligned} \tag{6.14}$$

$$\begin{aligned}\hat{\gamma}_2 - \hat{\gamma}_1 = ((p_{32} - p_{30} &- (\hat{\gamma}_1 - \hat{\gamma}_0)) \\ &+ (p_{42} - p_{40} - (\hat{\gamma}_1 - \hat{\gamma}_0)) \\ &+ (p_{52} - p_{51}))/3\end{aligned} \tag{6.15}$$

Here, $\hat{\gamma}_t - \hat{\gamma}_{t-1}$ is an average of all informative ways of estimating $\gamma_t - \gamma_{t-1}$ with individual properties taking as given all other index changes; there are four ways for $\gamma_1 - \gamma_0$ and three ways for $\gamma_2 - \gamma_1$. Note that if all properties had sold every period, then the only informative way to infer changes in index values $\gamma_t - \gamma_{t-1}$ from price changes of individual properties is to use successive sales of the same property. In this case, $\hat{\gamma}_t - \hat{\gamma}_{t-1}$ is just the average log price change of properties between those two dates. In this case, the index itself (not just the last normal equation) reduces to a log-linear version of (6.2), the conventional stock price index.

The normal equations (in either of the above forms) include index number values on the right-hand side; they are simultaneous equations. The solution of these simultaneous equations is (in terms of levels rather than differences of γ_t):

$$\hat{\gamma}_1 = .75\frac{(p_{11}-p_{10}) + (p_{21}-p_{20})}{2} + .25\left(\frac{(p_{32}-p_{30}) + (p_{42}-p_{40})}{2} - (p_{52}-p_{51})\right) \quad (6.16)$$

$$\hat{\gamma}_2 = .75\frac{(p_{32}-p_{30}) + (p_{42}-p_{40})}{2} + .25\left(\frac{(p_{11}-p_{10}) + (p_{21}-p_{20})}{2} + (p_{52}-p_{51})\right) \quad (6.17)$$

The estimate $\hat{\gamma}_t$ can be interpreted as a weighted average of all possible ways to infer the change in prices from 0 to t without using other index values, with weights that take account of the variance of the error of the inferred change. Since each of these equations is not taking as given the other index values, the variance of the error of the different ways of inferring γ_t is not the same for all ways. For both $\hat{\gamma}_1$ and $\hat{\gamma}_2$ above, the better way is multiplied by 0.75, the worse way is multiplied by 0.25, reflecting the fact that the variance of the error for the worse way is, under the ordinary least-squares assumption, three times as large as that of the better way. There are, at time 2, two different ways of estimating the price change between periods 0 and 1. The first way is just to look at the average price change of all properties sold in periods 0 and 1. The variance of this estimate here is one-half the variance of the regression error term, since this estimate averages two sale pairs. The second way is to take the average price change of all properties sold in periods 0 and 2 and subtract from this the average price change of all properties sold in periods 1 and 2. The variance of this estimate is one-half the variance of the regression error, for the average of the two properties sold in periods 0 and 2, plus the variance of the error term, for the error in the single property here sold at time 1 and time 2. These two ways of estimating the price change from 0 to 1 are independent of each other, since different houses are used and the regression model assumes no dependency across houses.[9] Similarly, there are, at time 2, two different ways to infer the price change between time 0 and time 2. The first and better way is to take the average log price change for all properties that were sold both in time 0 and time 2. The second and worse way is to take the average log price change for properties sold in periods 0 and 1 and add to this the average log price change for all properties sold in periods 1 and 2.

This interpretation of the regression estimates makes clear the significance of the assumptions that give rise to the ordinary least-squares regression. The variance of any of the ways of inferring price change is done by simply counting the number of sales used to compose it, and has nothing to do with the interval between sales. The regression model assumption would mean that, in estimating the price change between 1992 and 1993, inferring the price change by subtracting the log price change of a house sold in 1943 and 1992 from the log price change of a house sold in 1943 and 1993 is just as good as inferring the price change by averaging the log price change of two houses each sold both in 1992 and 1993. Clearly, the latter method is better, since there must be much more noise in the 49-year and 50- year price changes.

The ordinary least-squares method proposed by Bailey *et al.* (1963) is the appropriate way, from the standpoint of minimizing variance of estimates, to estimate the regression model only if the error term in the original model (6.8) is a random-walk variety that changes only at sale dates; that is, an independent error is added to the price at the time of each sale. In fact, in our real estate example there is likely to be heteroskedasticity in the regression error terms, and this suggests that a generalized least-squares procedure should be used. Webb (1988) and Goetzmann (1990) proposed variants of this estimator where the variances along the diagonal of ω were proportional to the interval between sales, reflecting a random-walk assumption for the regression error term. Case and Shiller (1987) proposed an ω matrix where the errors are linear (but generally with non-zero intercept) in the interval between sales, reflecting a model in which the regression error term is a random walk plus a time-of-sale noise term. In our regression estimates, both the random-walk component and the time-of-sale noise term were statistically significant. The generalized least-squares estimator using this error term was called the weighted repeat sales (WRS) estimator by Case and Shiller (1987), and the interval-weighted geometric repeat sales (IGRS) estimator in Shiller (1991).

While, from the standpoint of the regression model and estimation efficiency, we will want to use some form of the above generalized least-squares estimates to produce an index, from the standpoint of robustness to departures from the assumptions of our regression model, we may not want to do this. The regression model assumes that price changes of all time intervals are all derived from the same expected true γ_t, but this may not be the case. Properties with a longer interval between sales may be of a different class, and may have different price paths. Down-weighting these longer-interval price changes may then under-represent these properties in the index. In this case, ordinary least squares might possibly be better, in terms of bias considerations, than generalized least squares. In this case, however, there may also be other better ways to handle the heterogeneity among properties, using hedonic variables in the hedonic repeated-measures equation, or with ways that will be discussed in Chapter 7 below.

The normal equations for the IGRS (WRS) index $\hat{\gamma}$ are:

$$\hat{\gamma}_1 = \frac{w_{11}p_{11} + w_{21}p_{21} + w_{52}p_{51}}{w_{11} + w_{21} + w_{52}} - \frac{w_{11}p_{10} + w_{21}p_{20} + w_{52}(p_{52} - \hat{\gamma}_2)}{w_{11} + w_{21} + w_{52}} \tag{6.18}$$

$$\hat{\gamma}_2 = \frac{w_{32}p_{32} + w_{42}p_{42} + w_{52}p_{52}}{w_{32} + w_{42} + w_{52}} - \frac{w_{32}p_{30} + w_{42}p_{40} + w_{52}(p_{51} - \hat{\gamma}_1)}{w_{32} + w_{42} + w_{52}} \tag{6.19}$$

where w_{it} is the weight given to property i when it is the second sale of a repeat sales pair with second sale date at t. Thus, IGRS index values are differences of weighted averages of log prices. The weight is the inverse of the variance of the regression error term. In the indices of Webb (1988) and Goetzmann (1992), w_{it}^{-1} is proportional to the interval between sales. In the index of Case and Shiller (1987), w_{it}^{-1} is a constant plus a term proportional to the interval between sales.

Hedonic repeated-measures index

The ordinary repeated-measures indices just described included, in effect, hedonic variables, in the form of subject dummies, and in fact these dummies span all possible hedonic variables that are constant through time for each subject. It was argued above that we may want to include other hedonic variables in the analysis, as with the square feet of floor space variables shown in (6.8), to allow for time-varying effects of the hedonic variables on price. Moreover, we may have information that the quality of certain properties has changed through time. In estimating a real estate price index, for example, we may have information that the number of square feet in some homes has increased, due to the addition of extra rooms. Inclusion of hedonic variables may deal with criticisms of ordinary repeat sales indices that the sample of houses for which there are repeat sales data may be unrepresentative of all houses, Abraham (1990), Clapp *et al.* (1991).

The method of adding the hedonic variables illustrated in equation (6.8) preserves the repeated-measures experimental design. In contrast, other methods of using hedonic variables to produce repeat-sales indices (Case and Quigley, 1991; Case *et al.*, 1991) incorporate hedonic variables in such a way that the repeated-measures design is lost. In their methods, if there were a period of time when the mix of houses sold were changed in such a way that relatively higher-quality homes were sold, and if their hedonic measures failed to measure this increase in quality, then their price index would tend to show an increase even if all individual home prices had been constant through time.

Returning to the full example (6.8) above (not deleting the square foot dummies as in the preceding section), we are left with the geometric hedonic repeated-measures (geometric HRM) regression model $y = z\gamma + \varepsilon$, where z and y are defined by:

$$z = \begin{bmatrix} 1 & 0 & s_{11} & 0 \\ 1 & 0 & s_{21} & 0 \\ 0 & 1 & 0 & s_{32} \\ 0 & 1 & 0 & s_{42} \\ -1 & 1 & -s_{51} & s_{52} \end{bmatrix}, \quad y = \begin{bmatrix} p_{11} - p_{10} \\ p_{21} - p_{20} \\ p_{32} - p_{30} \\ p_{42} - p_{40} \\ p_{52} - p_{51} \end{bmatrix}. \tag{6.20}$$

A geometric hedonic repeated-measures index at time t could then be taken, after estimating the coefficient vector $\hat{\gamma}$, as the fitted value for a standard size house at time t: for example, a fixed-weight index at time t would be the coefficient of the constant term for time t (or zero if $t = 0$) plus $\bar{s}$ times the coefficient of the square foot variable for time t. Alternatively, suppose we have data for each time period t on the number of houses Q_{it} whose square feet of floor space falls in the ith of a number of intervals. One could then compute an index for the midpoint of each of these intervals, and use their antilog for P_{it} in (6.1) with these Q_{it}, to produce a value-weighted index. Doing this would not, however, produce a true value-weighted arithmetic index of house prices because individual houses are not treated arithmetically. Moreover, since our hedonic variables cannot describe completely the quality of each house, there are quality (and implied price) differences within each category, differences that may change through time.

In Chapter 7 below we will also use (6.20) with variables in place of hedonic variables s_{it}; s_{it} may be replaced by an estimated unobserved factor to proxy for unobserved hedonic variables, or by an inverse Mills ratio to correct for possible sample selection bias.

Arithmetic repeated-measures indices

Arithmetic variants of the above estimators can be obtained that show a closer analogy to equation (6.1), to the stock price indices, and to the consumer and producer price indices. The repeated-measures indices just discussed, using data on log prices, produce log indices whose antilogs are essentially geometric averages of prices. Arithmetic indices may be considered preferable, as noted above. The geometric indices have a portfolio interpretation that is lacking with the geometric indices. Portfolio values are arithmetic, not geometric, transformations of prices, since the value of a portfolio is the sum, not the product, of the values of the constituent assets. Using geometric averages to represent portfolio values introduces biases: for example, geometric averages are always less than arithmetic averages of the same positive numbers (so long as the numbers are not all the same).[10] Moreover, the geometric indices developed here are not value weighted; that is, in the real estate example, the more valuable properties do not get more weight, though they have more influence than less valuable properties in determining portfolio outcomes.

The importance of creating arithmetic value-weighted indices (like the stock price indices discussed in connection with expression (6.1) above) might be seen more clearly for art price indices. Changing tastes are likely to be more important factors in art prices than in real estate prices, since art serves more nearly a pure aesthetic function. Art of a style that falls out of favor may become of very little value, and yet, so long as the art continues to be sold, has just as much impact on a geometric art price index. Investors putting substantial sums into the art market will necessarily tend to put them into styles currently in favor; they do not want a hedging market to reflect average percentage changes in art prices, but average changes in portfolio value.

To go about creating value-weighted arithmetic indices, one is tempted first merely to replace the log prices in the above analysis with the prices themselves. But this would not do: the indices developed above were defined so that a log price is deflated to the base period by *subtracting* the index for that period, while we want our index to deflate the level of price by *dividing* by the index for that period. The value of the log index was 0 in the base period, and we want the value of the level index to be 1 (or 100) in the base period.

One must modify the above analysis by setting up a regression where some regression coefficients multiply the price; these coefficients are thus related to the *inverse* of the index (and in the case where there are no hedonic variables they equal the inverse of the index). Multiplying the price by the reciprocal of an index means deflating the price to the base year. But doing this requires that we turn the analysis around and make price an independent variable, rather than a dependent variable as it was in (6.8). Once we have prices as independent variables, we must in a regression model move to an instrumental-variables estimation approach, since we have stochastic variables (prices) among the independent variables.

Following a natural modification of our subject-dummy-augmented regression approach, we arrive at an arithmetic index that is derived by estimating a regression model $y = x\beta + \varepsilon$, a model to be estimated by instrumental variables using as the matrix of instruments the matrix z derived for the geometric index. Here, however, the vector y is a vector of zeros except for elements corresponding to a pair of sales where the first sale of the pair was in the base period, period 0, in which case the element of y is the price in the base period of that property. The x matrix is the same as the z derived above for the geometric index except that each element of the matrix is multiplied by the price corresponding to that property and date. For our simple example with $T = 2$, using capital P to denote levels of prices (in contrast to p which here denotes the natural log of prices), and where there are no hedonic variables, the matrix x and vector y are given by:

$$x = \begin{bmatrix} P_{11} & 0 \\ P_{21} & 0 \\ 0 & P_{32} \\ 0 & P_{42} \\ -P_{51} & P_{52} \end{bmatrix}, \quad y = \begin{bmatrix} P_{10} \\ P_{20} \\ P_{30} \\ P_{40} \\ 0 \end{bmatrix}. \tag{6.21}$$

Note that the row of this arithmetic ordinary repeated-measures regression model $y = x\beta + \varepsilon$ corresponding to sale pair i represents that the second price discounted using the index to the base period equals the first sale price discounted to the base period plus an error term. The regression model is estimated by instrumental variables, producing $\hat{\beta} = (z'\omega^{-1}x)^{-1}z'\omega^{-1}y$ where z is given in (6.10). Since there are no hedonic variables, the index (with base period, $t = 0$, value equal to 1.00) for time t is just $\hat{\beta}_t^{-1}$.

In the case where there is no heteroskedasticity correction, where the matrix ω is proportional to the identity matrix, the implied normal equations $(z'x)\hat{\beta} = z'y$ are:

$$\hat{\beta}_1^{-1} = \text{Index}_1 = \frac{P_{11} + P_{21} + P_{51}}{P_{10} + P_{20} + \hat{\beta}_2 P_{52}} \tag{6.22}$$

$$\hat{\beta}_2^{-1} = \text{Index}_2 = \frac{P_{32} + P_{42} + P_{52}}{P_{30} + P_{40} + \hat{\beta}_1 P_{51}} . \tag{6.23}$$

The index defined by equations (6.22) and (6.23) was called the Value-Weighted Arithmetic Repeat Sales price index (VWARS) (Shiller, 1991). Note that equation (6.23) is the same as equation (6.3), so that the analogy of this arithmetic index to the stock price indices is apparent. Note also that if each property is sold each period then the index value for all time periods is given exactly by the usual chain formula (6.2), the conventional stock price index. To see this in this example, delete properties 3 and 4, then substitute (6.23) into equation (6.22).

The same arithmetic index number construction procedure, if account is taken of the heteroskedasticity of residuals (ω is diagonal but not proportional to the identity matrix), gives us instead the normal equations:

$$\hat{\beta}_1^{-1} = \text{Index}_1 = \frac{w_{11}P_{11} + w_{21}P_{21} + w_{52}P_{51}}{w_{11}P_{10} + w_{21}P_{20} + w_{52}\hat{\beta}_2 P_{52}} \tag{6.24}$$

$$\hat{\beta}_2^{-1} = \text{Index}_2 = \frac{w_{32}P_{32} + w_{42}P_{42} + w_{52}P_{52}}{w_{32}P_{30} + w_{42}P_{40} + w_{52}\hat{\beta}_1 P_{51}} \tag{6.25}$$

where the weights w_{it} are as defined in connection with expression (6.19) above. In the case where the heteroskedasticity correction represents error variance in terms of the interval between sales, as in Case and Shiller (1987), the index was called the IVWARS (Interval and Value-Weighted Arithmetic Repeat Sales) index (Shiller, 1991). Of course, making a heteroskedasticity correction means that the index is no longer strictly value weighted, and may not represent the value-weighted portfolio if there is a correlation between the variance assumed for the heteroskedasticity correction and the value of the properties. If the basis for the heteroskedasticity correction is the relation of variance to the interval between sales, then this correlation may be low.

We can also include additional hedonic variables in the analysis, to produce arithmetic hedonic repeated-measures indices. Following the same argument as above, we produce the matrix x by multiplying each element of the z matrix corresponding to subject i at time t by P_{it}. Taking s_{it} as the square feet of floor space, we would begin estimating the arithmetic hedonic repeated-measures (arithmetic HRM) regression model $y = x\beta + \varepsilon$ by $\hat{\beta} = (z'\omega^{-1}x)^{-1}z'\omega^{-1}y$ where z is as in (6.20) and x and y are given by:

$$x = \begin{bmatrix} P_{11} & 0 & P_{11}s_{11} & 0 \\ P_{21} & 0 & P_{21}s_{21} & 0 \\ 0 & P_{32} & 0 & P_{32}s_{32} \\ 0 & P_{42} & 0 & P_{42}s_{42} \\ -P_{51} & P_{52} & -P_{51}s_{51} & P_{52}s_{52} \end{bmatrix}, \quad y = \begin{bmatrix} P_{10} \\ P_{20} \\ P_{30} \\ P_{40} \\ 0 \end{bmatrix}. \tag{6.26}$$

Then a fixed-weight arithmetic hedonic repeated-measures index for a standard property with $\bar{s}$ square feet of floor space for time t can be taken as the inverse of the sum of the estimated coefficient of the price variable corresponding to time t and $\bar{s}$ times the estimated coefficient of s_{it} for that time period.[11] To get a chain index, one could first derive index values for each of various categories defined by square feet, then substitute these along with quantities outstanding in each category into equation (6.1).

Note that in this example it may be desirable to impose a heteroskedasticity correction that models the error variance in terms of the quality variable s_{it} as well as the interval between sales. Note also that if we used for s_{it} in x, not square feet, but a dummy variable that is 1 if property i is in a certain category of properties and is 0 otherwise, then the coefficients of the first two columns will correspond to (inverse) price indices constructed using (6.21) for properties not in this category; the coefficients of the second two columns will correspond to the difference between the (inverse) price indices constructed using (6.21) for properties in this category and not in this category. However, the fixed-weight hedonic index for time t taken as the inverse of the sum of the coefficient of the price variable corresponding to time t and $\bar{s}$ times the coefficient of $P_{it}S_{it}$ for that time period for some fixed proportion $\bar{s}$, $0 < \bar{s} < 1$, is a weighted harmonic average of the two indices, not the weighted arithmetic average that would be suggested by (6.1).

As with (6.20), we might consider methods that replace s_{it} in (6.26) with an estimated factor to represent an unobserved quality variable, or by an inverse Mills ratio to correct for sample selection bias; see Chapter 7.

7

Index Numbers: Issues and Alternatives

Creating index numbers for settlement of contracts requires some judgment; no single method is likely to be applicable to all circumstances. There are tradeoffs, and choices have to be made with limited information. Before applying a repeated-measures method like the ones defined in the preceding chapter, we have to decide whether there are enough repeated measures to ensure that the standard errors are not going to be too high. We have to believe that there is enough unmeasured quality variation across subjects to warrant the increase in error variances caused by the addition of many subject dummies.

We must choose which kinds of hedonic variables, if any, to include in the analysis. Not all quality measures are appropriate for index number construction. We must choose whether these variables or the subject dummies are to be constrained in any of various ways. Prior information of an imprecise nature may be used to put probabilistic, rather than rigid, restrictions on the regression coefficients.

There are also some fundamentally different variants of the hedonic repeated-measures regression methods that could be considered, methods in which quality is inferred as an observed factor associated with each subject, and in which a separate selection equation is used to correct for possible selection bias in the mechanism by which it is determined which subjects are to be measured.

Expected incidence of repeated measures: standard errors

The decision whether to use a repeated-measures index, rather than a hedonic regression index that does not include subject dummies, depends ultimately on whether there will be enough repeat-sales pairs so that the method does not result in unacceptably high standard errors. That is, the decision depends on whether the time interval over which data are observed is long enough, given the probability of a sale in any time period, and, given that, the sample size is large enough, to produce a sufficient number of repeat-sales pairs so that the standard errors of the regression estimates will be low. Thus a critical factor in the decision is how far into history the data extend; we need substantial historical data as well as current information to produce repeated-measures indices.

There is some concern whether we can expect, given the time intervals spanned by existing historical data sets, to find enough repeated measures to make worthwhile the application of such indices. In Case and Shiller (1987), which used a data set on single-family homes sales from 1970 to 1986 in four cities, the total number of repeat-sales pairs divided by the total number of sales was only 4.1%. This finding suggests that much may be lost in terms of standard errors by using a repeated-measures regression approach. But that data set did not contain the universe of sales in a given city in the time interval covered, only a sampling of sales there. The probability that a given house sold twice is also sampled on both sales may be small. What is the likely incidence of repeat-sales pairs when we have data on the universe of sales in a given region or class of properties in a given time interval?

Binomial model

It is attractive to assume, as a first approximation, that the number of times a given subject is measured (property is sold) in an interval of time has a binomial distribution with probability p of a sale in any unit of time. In our real estate application, many risks that cause a property to be sold are random events. House sales, for example, may be caused by such things as job changes, deaths, divorces, or changes in wealth or credit status. The binomial model will work poorly for houses if houses are sold, let us say, primarily on life-cycle reasons, e.g. everyone owns a single house during the childbearing years.

If a property is sold x times, then, if $x > 1$, there are $x - 1$ repeat-sale pairs, otherwise there is no repeat sale. The expected number of repeat sales in a sample from 0 to $T - 1$ (T periods) for a given property is:

$$E\,(\text{No. of repeat sales}) = \text{prob (2 sales)} + 2 \times \text{prob (3 sales)} + \cdots + (T-1) \times \text{prob } (T \text{ sales}). \tag{7.1}$$

Under the binomial assumption this expectation is:

$$E\,(\text{No. of repeated measures per property}) = \text{mean of binomial} - 1 + \text{prob (0 sales)} = pT - 1 + (1-p)^T . \tag{7.2}$$

This expectation can of course exceed 1.00.

If the number of houses from which the sample is derived is large, the expectation of the ratio of (pairs of) repeat sales to all sales is approximately the ratio R of the expectations:

$$R = \frac{pT - 1 + (1-p)^T}{pT} . \tag{7.3}$$

The ratio cannot exceed 1.00; there cannot be more pairs of repeat sales than there are sales; the ratio approaches 1 as T goes to infinity. With a large sample, each property is sold many times; if it is sold k times then the ratio for that property of pairs of repeat sales to sales is $(k-1)/k$; the limit of expression (7.3) as p goes to 1.00 is $(T-1)/T$.

Fig. 7.1 shows R for various values of T (years of sample length). The curves are shown for three values of p, the probability that a given asset sells in one period of time. Thus, for example, if we have a sample of 100,000 house sales, comprising all sales in a stable community over a 20-year interval, 5,000 sales occurring per year, and if the probability that a given house sells in a given year is always 0.1, then we would, by this model, expect to find (reading from the figure) 56,000 repeat-sales pairs.

In data sets that show all sales of a fixed group of houses in an interval of time, to get R as low as 4.1% in a sample of 66 quarters, as in Case and Shiller (1987), would require that the probability of selling a house in a quarter is only .0013, or about half a per cent per year; in fact, about 6% of houses sell in a year in the United States.[1] Clearly, the Case–Shiller (1987) data were

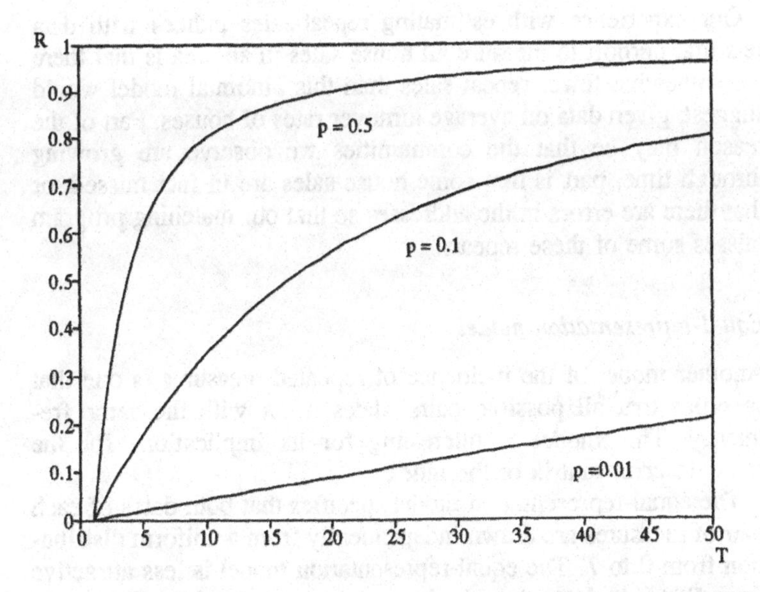

FIG. 7.3. The ratio R of expected pairs of repeat sales to expected all sales in a sample of T time periods, for various p, probability of a sale, expression (7.3).

not in accordance with this model, presumably because they did not represent all sales of a fixed population of houses. If we add to the above binomial story that the probability that a given house sale is reported in our data is s, and if the event of reporting the sale is independent across sales and houses, and if the number of houses from which the sample is derived is large, then the expectation of the ratio of (pairs of) repeat sales to all sales is approximately given by the ratio R above where sp replaces p. Then, the ratio R of 4.1% observed in the Case–Shiller (1987) data described here would be consistent with a roughly 6% probability of sale per year if s is of the order of 10%.[2]

Note that, with the binomial assumption, sales that are members of one or more repeat-sales pairs occur with the same expected frequency across all periods of the sample, but there will tend to be relatively few second sales near the beginning of the sample, relatively few first sales near the end. If T is large, the distribution of repeat-sales pairs by interval k between sales ($k > 0$) is approximately geometric, $(1 - p)p^{(k-1)}$.

Our experience with estimating repeat-sales indices with data sets that purport to measure all house sales in an area is that there are somewhat fewer repeat sales than this binomial model would suggest, given data on average turnover rates of houses. Part of the reason may be that the communities we observe are growing through time, part is that some house sales are in fact missed, or that there are errors in the addresses so that our matching program misses some of these repeats.

Equal-representation model

Another model of the incidence of repeated measures is one that specifies that all possible pairs' dates occur with the same frequency. This model is interesting for its implications for the standard error matrix of the index.

The equal-representation model specifies that both dates of each pair of measures are drawn independently from a uniform distribution from 0 to T. The equal-representation model is less attractive from first principles than the binomial model, but it may serve as a reasonable approximation to the binomial model if the expected number of measures per subject pT is not too large; this can happen if the historical sample interval T is not too large or the probability p of a measure (sale) is not too large. Then the binomial distribution implies that there are very few subjects measured more than two times; our sample of observations will tend to show each subject represented as measured twice. (There are, of course, many more subjects measured only once or measured not at all, but these subjects are deleted from our sample in the procedures that gave rise to the repeated-measures indices described above.) To the extent that subjects are measured only twice, then, since the binomial model implies equal probabilities of a measure at each time period, the equal-representation model applies.

There are $(T^2 - T)/2$ possible kinds of repeat-sales pairs in terms of dates; for example, in expression (6.10) the matrix z, for the case $T = 3$, shows all three kinds of repeated-measures pairs; in that matrix, the first and second row show one kind, the third and fourth row show another, and the last row shows the third. In that z matrix, the three kinds do not appear in the same proportion, since there is only one row corresponding to the third kind.

This equal representation model implies a frequency distribution from 1 to $T - 1$ of intervals between dates of repeated-measures pairs in the sample that declines linearly with the interval.

If all kinds of pairs do occur in the same proportion as assumed with this model, and supposing that the number n of repeat-sales pairs is a multiple of $(T^2 - T)/2$, then the matrix $z'z$ has all of its diagonal elements equal to $2n/T$ and all of its off-diagonal elements equal to $-2n/(T^2 - T)$. The inverse of this matrix has all of its diagonal elements equal to $(T - 1)/n$ and all of its off-diagonal elements equal to $(T - 1)/(2n)$; the standard error of an index value is therefore equal to the standard deviation of the error for one house divided by the square root of the number of repeat-sales pairs per index value estimated. One may think of this variance matrix as the variance matrix one would find for the index values for times 1 through $T - 1$ if the index were derived from estimates of price for times 0 through $T - 1$ that had independent, identically distributed errors. Since time 0 is, by our normalization, the base period, the (log) index would be, in this interpretation, computed by subtracting the time 0 error from the errors for times 1 through to $T - 1$. Thus, the error, for any time period from 1 through to $T - 1$, has two components to it: its own component (which is uncorrelated with anything else) and a component due to the normalization (a component which is shared completely by every other error).

With this variance matrix for the errors, time series plots of the index would tend to appear contaminated by white noise. This is what we often find with actual residential real estate price index data. When these indices are produced using a lot of data, so that the standard errors of the index are small, then plots of the estimated indices often look like rather smooth curves (since underlying real estate prices show substantial persistence through time) drawn with a shaky hand or an inaccurate plotter, producing many erratic little wiggles. The variance matrix also implies that errors in the period-to-period growth rates (the first differences of the log index) behave exactly like first differences of white noise, since the component of error that is shared by all index values is eliminated by differencing. Thus, in small samples, so that standard errors are large, time series plots of growth rates in the indices tend to have an irregular, saw-toothed appearance, the period-to-period growth rates frequently alternating in sign.

This variance matrix has the same structure as the variance matrix one would have if the index were derived as a simple average of independently sampled (log) prices each period, and where the sample size and variance of price dispersion is the same each period. To the extent that consumer price index computations and error variances can be described in this way, the variance matrix structure for the repeated-measures index is the same as for the simple consumer price index. This may seem surprising, since the repeated-measures index is derived from observations of overlapping intervals.

The variance matrix for errors in this repeated-measures index has the unusual implication that standard errors of coefficient (log index) differences over any differencing intervals are the same; they are always equal to the standard deviation of the error for one house divided by the square root of the number of repeat-sales pairs per time period. For example, the standard error of the change in the log index over an interval of ten years is the same as the standard error of the log index over one month. One might have thought that standard errors of index changes over longer differencing intervals might be larger; indeed, the index changes tend to be larger and more variable over larger intervals.

Of course, this equal-standard-error result depends on the assumption of ordinary least squares that the variance of the regression error term is constant, and hence that it does not depend on the time interval between sales. In practice, longer-time-interval repeat-sales pairs are likely to have larger errors in the regression, because there is more time for the property to show a deviation from the aggregate price performance. If a generalized least-squares regression method (such as the IGRS (WRS) method, as in (6.18) and (6.19) above) is used to correct for these unequal error variances, in effect downweighting the longer-time-interval repeat-sales pairs in the regression, then there will be a tendency for standard errors to be larger on longer-interval index differences. Even so, the effect of such downweighting in causing unequal standard errors may be much less than one might think. If the equal-representation model holds, the standard errors on two-period price changes will be increased, by a factor less than two relative to the standard errors of one-period price changes, by using generalized least squares based on the assumption that the standard deviation of the error in the two-period-apart repeat-sales pairs is twice that of the one-period-apart repeat-sales pairs. The reason is that the two-period changes in index values are derived in part from the information in the one-period-apart repeat-sales pairs that are not downweighted, and the one-period changes in index values are derived in part from the information in the two-period-apart repeat-sales pairs that are downweighted.

In applied work with real estate price data, one often finds that there is no pronounced trend in the standard errors through time or as the differencing interval is increased. Such was the finding in the IGRS (WRS) repeat-sales log price indices produced in Case and Shiller (1987) for four US cities 1970–86, even though a heteroskedasticity correction was employed that downweighted repeat-sales pairs with a longer-time interval between sales.

Restrictions on coefficients of subject dummies and hedonic variables

The high standard errors that may result when hedonic repeated-measures indices use up too many degrees of freedom given the sample size can be dealt with by restricting the regression coefficients corresponding to subject dummies or hedonic variables.

It is not always desirable to include all subject dummies in an unrestricted form, as in the last chapter, in a hedonic regression equation. Subject dummies are useful only if there is some reason to expect that they have sufficiently non-zero coefficients, that is, that there is an element of quality that is associated with subjects and not represented among other included hedonic variables. This is not to say that quality is absolutely constant for subjects, only that there is a component of quality that is constant. If, however, there is no stability to the quality of subjects—their characteristics are constantly and totally changing—then the subject dummies will be redundant variables. Redundant variables, variables with zero coefficients, do not affect the unbiasedness property of linear regression, but they do increase the standard errors of the coefficients. When we are adding subject dummies, we are adding a lot of extra variables, with significant potential effects on efficiency of estimation.

It is possible that in some circumstances one may wish to include a smaller set of dummies than the set of subject dummies, call them class dummies, which represent the class or kind of subjects, classes in which subjects are expected to have fairly constant quality differences unmeasured by the included hedonic variables. The other subjects would remain as single-sales observations in the regression. Doing this selectively would be justified by the idea that single-sales observations are included only when there is reason to believe that the included single sales represent properties whose enduring characteristics are thought to be very well measured by the included hedonic variables. Doing this would mean ignoring the other single-sales data.

It may be a worthwhile assumption that all condominiums in a particular building, or all condominiums in a particular building that face in a certain direction, may share a single class dummy. Doing this may be especially worthwhile if there are other hedonic variables (such as square feet of floor space) available, that might correct for individual differences across the condominiums. In estimating an index of aircraft or railway equipment prices, one may wish to assign subject dummies not to individuals but to individual model numbers.

One common restriction that is placed on the regression is to constrain the quality variables other than the constant term to have the same coefficient through time, which yields what may be called a time-dummy hedonic regression. For example, we might modify the giant regression (6.5),

$$Y = \begin{bmatrix} Y_0 \\ Y_1 \\ Y_2 \end{bmatrix}, \quad Z = \begin{bmatrix} 1_0 & 0 & 0 & Z_0 \\ 0 & 1_1 & 0 & Z_1 \\ 0 & 0 & 1_2 & Z_2 \end{bmatrix} \tag{7.4}$$

where 1_t is an n_t element column vector of 1s, and the Z_t are now $N_t \times K - 1$ element matrices with the constant terms deleted. There are many variations on this theme: we could constrain some but not all of the quality variables to have constant coefficients through time, or we could allow the coefficients of quality variables to change through time, but apply some Bayesian or ridge-regression smoothing to their changes. (Such Bayesian methods will be discussed below.) Any of these methods could be used to construct Laspeyres, chain, or other indices, as above. But when subject dummies are appended to the regression as in the last chapter, any hedonic variable that is so constrained will drop out of the hedonic repeated-measures regression, and will result in a column of zeros in the z matrix that will have to be deleted, unless the quality varies through time for at least one subject. Restricting the coefficients of the hedonic variables in this way thus means, in a repeated-measures context, that use will be made of the hedonic variables only to correct for changes in these variables, e.g. to correct for the effect of enlargements of houses by adding a single variable representing the change in square feet of houses between sale to the ordinary repeated-measures regression.

Some hedonic price index construction methods involve constraining all coefficients, including the constant terms, to be the same through time. Such methods would in effect create a Z matrix of dimension $K \times N$ by stacking the matrices Z_0 through Z_T on top of one another, so that in the model $Y = Z\gamma + \varepsilon$ the vector γ has only K elements. Obviously, creating an index from such an estimated regression, by taking the fitted value of the regression for a standard subject (or taking a chain index from the regression), will produce an index that does not change through time. However, when such a regression is run to construct a price index, the index is taken not as the fitted value for a standard, unchanging subject but instead as the fitted value in terms of some time series of independent variables. For example, to construct a real estate price index, one may include the rent on the property in the hedonic regression as the independent variable. Its estimated coefficient is then taken as an indication of the rate at which rents are discounted into today's price. The index at time t is taken as the fitted value of the regression for the rent observed at time t. Such a model is based on a maintained hypothesis that prices must be determined by rents and does not allow any other variation in prices. Such price indices would seem to be derived from such restrictive assumptions that they cannot really be regarded as genuine price indices; in this case they would disregard all evidence of a real estate price boom if the rents did not also boom. They take no account of the possibility that prices incorporate information not in our information set, information, say, about future rents rather than today's rent, nor do they take into account any variation in prices that is purely speculative.

Smoothing and improving indices: Bayesian methods

It may be useful to replace the above classical regression methods for estimating indices with Bayesian methods, which put some informative prior on the change in the price level to be estimated, on the coefficients of hedonic variables, or on the coefficients of the subject dummies.[3]

Since historically we know that changes in real estate prices have some central tendency, that, for example, month-to-month price changes exceeding 5% are quite rare, we may wish to put an informative prior on the price changes that incorporates this information into our estimation procedure. The use of such priors would tend to smooth out the estimates, reduce anomalous observations of sudden price change, index values that might be due to nothing more than a sparsity of observations for the time period of that observation. Moreover, we know (see Chapter 5) that, historically, changes in real estate price indices have been serially correlated: periods of increasing prices tend to be followed by periods of increasing prices; periods of decreasing prices by periods of decreasing prices. The use of priors representing this information would tend also to smooth out the estimates, eliminating the saw-toothed pattern of errors in growth rates described above in connection with the equal representation model. Bayesian priors might even exploit the known historical relation between changes in real estate prices and changes in other economic indicators to produce an even better estimate of price.

Those who trade in the index may wish to see the index improved as much as possible, even if the methods used to effect the improvement are, at their foundation, somewhat judgmental. Using formal Bayesian priors as part of an estimation procedure is certainly no more judgmental than are other futures market settlement techniques, certainly less so than is the procedure of tasting coffee referred to above.

On the other hand, the use of such informative prior information in the construction of the index reduces the simplicity and objectivity of the index construction method. Such Bayesian or ridge methods do amount to artificially smoothing or otherwise doctoring the data, and we may feel some resistance to doing so. How exactly, after all, do we smooth the data? How much do we reduce short-run fluctuations, and what do we assume about the ultimate smoothness of the true price level? How much serial correlation in prices should our priors assume? Should we assume that the pattern of serial correlation of price changes is constant through time, or should we devise a method that responds gradually to changing stochastic properties through time? These are questions that people will differ on, and there is no objective answer.

If we were in the business of producing index numbers solely for the purpose of publishing the results as a sort of 'leading indicator' of housing market conditions, then we might well say that no smoothing should be done at all. If the index numbers were to be used only as leading indicators, then people might well object to an index number construction method that, in smoothing the data, produces an estimated index that shows little change from the preceding period even if all sales in that period showed a large change; this index may be seen as doctoring the truth. People may well wish to have data presented to them in as unfiltered a manner as possible, so that they themselves can judgmentally adjust it as they see fit. Most of the major government-produced economic statistics impose no smoothing, although seasonal adjustment is commonplace.

On the other hand, the argument that the evidence should 'speak for itself' and not be filtered to reflect the priors of the data constructors disregards the fact that it may be difficult for data constructors to represent their information adequately to the consuming public. The public may not understand standard errors, for example. The index number publisher may well know that an apparent jump in the index is accompanied by a large jump in the standard error of the index (as may happen with real estate if there are few sales of properties in that period.) Rather than try to explain this matter to the public, it may serve the public's interests better if the index is simply smoothed to eliminate the jump. Moreover, if the index number construction method is to be used for cash settlement, then people might want to see some sort of smoothing imposed, so that they will not find themselves settling, and thereby gaining or losing large sums of money, on an observation that is suspected to be anomalous.

Ultimately, the conclusion whether to use such Bayesian methods hinges on the quality of the data. If there are abundant data, then we may well feel that there is no point in introducing complexities in the index construction method. If the data are very sparse or filled with errors, then we may be practically forced to use some smoothing method in construction of an index.

Choice of hedonic variables

The hedonic repeated-measures regressions described in the preceding chapter can be used to construct indices that control well for changes in the mix of subjects measured. To do this, constructors of indices should not shrink from using many hedonic variables, should these variables be available. For example, in constructing art price indices, it was noted in Chapter 5 that there is a tendency for sales of paintings by artists currently in vogue to sell repeatedly and to show larger price movements than are occurring with prices of painting by artists not in vogue. But artists are individual people, and there is a vast number of them. In constructing art price indices, then, there is a very serious problem of change in mix. We might want to include a dummy variable, among the hedonic variables, for each artist. The price indices derived for individual artists may then have extremely high standard errors; but an aggregate price index, that gives fixed weight (in each point of the chain) for each artist, need not have high standard errors, since it averages many artists. Such an index would eliminate the problem, noted above, that many Picassos may be sold in certain years at rapidly inflated prices; Picasso would get only his just weight corresponding to his share of art in collections. Of course, if a dummy is included for every artist, then there will surely be a problem of multicollinearity (there will be columns of zeros in the matrix Z_A defined in the preceding chapter). Dummy variables might be created that group some minor artists together, and retain a Picasso dummy if Picasso is sold regularly. Bayesian methods that represent some probabilistic relation among artists' prices may also break the multicollinearity problem, and may better represent our prior information about art prices. Judgment is, of course, involved in grouping artists together or specifying their similarity; this is the same sort of analysis that Sotheby's art appraisers apply to compiling appraisal-based indices of art prices. Hedonic variables, in index number construction for contract settlement of the kind envisioned here, must normally be variables that identify a particular claim on a cash flow.

Since the index number construction methods here are intended to give us a representation of the change in price of a standard claim on income or services, we would not want to include a time-varying hedonic variable that represents changing income of the claim. For example, we would not want to include the time-varying rental rate on the property among the hedonic variables. Taking an index equal to a fitted value of a standard property would mean making an index number of the price of a property of a standard rent. Since the rent on individual properties changes through time, this would not be an index of the changing price on existing properties. We should exclude the rent from the regression even if that variable is statistically significant when it is included. The hedonic regression is to be regarded merely as a projection of prices onto characteristics of the investment, and must exclude some variables that might improve fit. By the same token, we would want to leave out of the list of independent variables a time-varying measure of neighborhood quality or of crime rates in the area.

In constructing an index to serve as the basis of settlement in derivative markets, we must include only independent variables that identify the investment. We must leave out of the list of independent variables any variables that naturally change through time for existing properties, because these may be part of the process by which prices and incomes of existing properties change. There are, however, a couple of exceptions to this rule. First, we could include variables representing new investments in the list of independent variables that change through time, such as changes in the number of square feet of floor space. Second, we could include a hedonic variable that, while the variable does change through time, is entered into the regression in such a way that it does not change between measures. That is, we could include in our z and x matrices the rental rate on the property at first sale in the columns representing both the first and second sales. We could then develop indices that represent the behavior of portfolios of real estate that were high rent or low rent, say, at the date of initial investments in the property.

Another time-varying variable that could be kept constant between measures in this way for the construction of real estate price indices is assessed value. Clapp and Giaccotto (1992*a*; 1992*b*) have advocated using assessed value in place of first sale price, to run a repeat-sales-like regression without data on actual repeat sales. The assessed value could also be used in a hedonic repeated-measures regression if assessed value at first sale is used for both sale dates, as described above. If there are two assessed values, values assessed at two different assessment dates, then both assessed values could be included as separate hedonic variables, on the theory that they are two independent measures of quality. Assessed values may be attractive to include among the hedonic variables, since they are available for every property and represent attempts at direct measurement of the quality of the property.

Hedonic variables related to improvements

Accounting for improvements made in the properties for which we are producing price indices could in principle be very important. Price indices could show an increase that is not due to any change in market conditions. If many homeowners add new rooms to their houses, for example, a price index that does not account for these improvements will show an increase. Whether we want to allow such index increases in response to improvements depends on the hedging function the indices will serve. For individual homeowners hedging the value of their homes in futures markets, the improvement-generated index increases would be undesirable for contract settlement, since the increases would cause shorts to pay at settlement even when there was no change in the return on their housing investment. On the other hand, for holders of mortgage-backed securities that are using a residential house price futures market to hedge the risk of prepayment or default, the improvement-generated index increases would be desirable; for them it is immaterial whether the price increases were generated by homeowner expenditures.

Should we decide that we do wish to remove the effect of improvements on the price index, the hedonic repeated-measures index described above lends itself naturally to including hedonic variables representing changes in the quality or size of the assets traded. In the real estate example featured above, properties that have been improved by additions will automatically be taken account of by the changes in the square-feet variable between columns of the z or x matrices defined above.

Many data sets do not, however, include regular time-series data on characteristics of individual properties sold such as the square-feet variable used in the example here. We might also consider including as hedonic variables time series data on total expenditures on improvements for the geographical region in which a house is situated, so long as there is more than one such region within the geographical area for which the index will be computed, as described above. The coefficient of these variables might be interpreted as soaking up the effects of the improvements on market price, allowing us to compute a price index that is not biased by improvements.

It should be noted that economic theory does not reliably tell us much about the effect of improvements on price; and we do not always know even the direction of the effect. We do not know how much expenditures made to a given property by its owner will enhance (or even detract from) its market value when it is later sold. Many expenditures made by homeowners are done to change the property to their own tastes and needs; these expenditures may not be valued highly, or at all, by subsequent owners.

Trying to take account of the effects of changed properties by using any such data on improvements runs some risks, and we may instead wish merely to exclude changed properties from the regression or ignore the changes altogether, that is, use s_{it} which is made to be a constant through time. There may be a spurious correlation between improvements and price changes, if there is feedback from price changes to improvements. Suppose, for example, that homeowners are encouraged to increase the size of their homes in response to increasing market prices. There might then be a correlation between improvements and market prices even if improvements have no effect on market prices. The estimated regression coefficients might imply an exaggerated effect of improvements because of this spurious correlation, thereby causing biases in the index constructed using the remaining coefficients.

Fortunately, it does not appear that the price uncertainty that we observe in residential real estate markets has much to do with uncertainty about the pace of improvements. In the United States, the value of residential new construction improvements averaged only $53.0 billion per year from 1984 to 1990, less than 1% of the value of residential real estate. And of this, most was for alterations and major replacements, for which the impact on home value would seem likely to be substantially less than one-to-one. The average value of residential structure additions in the United States from 1984 to 1990 was only $7.7 billion per year, roughly a tenth of 1% of residential real estate value. Moreover, even this expenditure on structure additions was fairly steady, so that its variations would not contribute much to year-to-year variation in price indices. The standard deviation from 1984 to 1990 of expenditures on structure additions was only $2.2 billion, or less than four one-hundredths of 1% of the value of residential real estate.[4]

Hedonic variables related to depreciation

Indices of price of assets that depreciate through time would tend to show a down-trend through time reflecting this depreciation. It was noted on Chapter 5 above that this down-trend might be a striking difference between repeat-sales indices and conventional indices, such as components of the producer price indices of durable goods. The latter are prices of new goods.

There has been substantial discussion of how to infer depreciation measures from data on selling prices. Cagan (1971) and Hall (1971) discussed doing this using vintage price data on same models of consumer durables. Cagan was able to find models of automobiles that, according to *Consumer Reports* magazine, had been essentially unchanged from the previous year. The relative price between new and used models was taken to be a measure of depreciation. This data set is, however, unusual; rarely do we have data indicating identity of subjects of different vintages.

Chinloy (1977) proposed a method of estimating depreciation using a framework related to the Bailey *et al.* (1963) repeat-sales method. His method was criticized by Palmquist (1979). Palmquist argued that the essence of Chinloy's method was adding the interval between sales as an extra hedonic variable to the repeat-sales regression (i.e. to a regression using matrices like those exemplified by (6.10)). Palmquist noted that if this had been done properly there would have been strict multicollinearity between the time-interval-between-sales variable and the dummy variables in the z matrix. The summation of t times the tth column of z, $t = 1, ..., T - 1$, equals the time interval between sales.

Indeed, it appears that Chinloy's concept of depreciation was rather different from that we usually accept; there should be no way to decipher from price data alone whether price declines are due to depreciation or decline in market value. We could, on the other hand, include a nonlinear transformation of the interval between sales. Suppose we think that many houses have new kitchens installed immediately after they are sold, and that these new kitchens depreciate mostly in a few years. Then a variable $e^{-rm_{it}}$, where r is an assumed depreciation rate and m_{it} is the interval between sales for property i sold for the second time at time t, could be included in place of s_{it} in our hedonic repeated-measures regression.[5] If we have reason to think that, say, most depreciation occurs in the first few years after a property is built, then we could instead include in the regression a non-linear transformation of the age of the property. This might allow us to correct in some measure for depreciation, but the results would be sensitive to the functional form assumed.

As noted above, for the purpose of creating indices for contract settlement, there would appear to be no reason to try to purge a price index of effects of depreciation. To the extent that depreciation is non-stochastic, it has no effects on the hedging ability afforded by derivative markets. To the extent that it is stochastic, then it represents another risk that futures markets can be used to hedge. Still, introducing age of subject as a hedonic variable in a hedonic repeated-measures equation may be useful for the purpose of creating indices for contract settlement because it can help us to control for a change in mix (in terms of age) among properties observed. Without a nonlinear transformation of the age variable we could not, because of multicollinearity problems, include all of the columns corresponding to age. We could not, for example, add all columns including s_{i0} in equation (6.20) if s_{it} were age of property; the sum of these columns would be the interval between sales. But we could include all but one of them, by running the regression exemplified by (6.20). Including columns for such an age variable and producing an index for properties of a standard age would prevent the settlement of contracts from inaccurately representing price changes if, for example, a lot of older properties happened to sell at the time of the settlement date, and if their price paths were not representative of all prices.

Market conditions as hedonic variables

While we normally expect to use hedonic variables only as identifiers of specific claims on future income or services, there might be an exception to this rule; one might sometimes wish to use hedonic variables that represent market conditions. In developing price indices for assets that trade on sluggish, illiquid markets, the question naturally arises whether we should include among the hedonic variables factors such as vacancy rates or average time to sell, and thereby correct for these conditions. This question leads us into some definitional problems, about just what it is that we want to measure.

In constructing price indices we seek some measure of 'market value', but the definition of this term is ambiguous. According to one survey of appraisers' definitions of market value:

> The term 'market value' is understood in many ways: as a symbol, norm, opinion, inference, expectation, prediction, event, ideal. Out of this thicket of meanings no unifying idea has emerged despite years of effort by the appraisal profession and the courts. The official definitions are clumsy, lax, unrealistic, and self-contradictory, obviously the work of tired committees and overworked jurists. If there is a single concept common to all of them, it has yet to be articulated.[6]

The most common definition of market value for real estate, which was stated in the case of *Sacramento etc. R. R. Co. v. V. Heilbron*, is:

> The highest price estimated in terms of money which the land would bring if exposed for sale in the open market, with reasonable time allowed in which to find a purchaser, buying with knowledge of all the uses and purposes to which it was capable of being used.[7]

This definition is unfortunately not very precise. There is great latitude for interpretation of the meanings of the terms 'highest price', 'open market', 'reasonable time', and 'knowledge of all uses and purposes'. Consider the phrase 'reasonable time'. Certainly, the price that a seller can expect to get for a property rises monotonically with the time the seller can allow to keep it on the market. If the seller must sell today, the price will be very low. It will be a distress sale to whomever is willing to take the property today just because the price is low. If the seller has a month to sell, there may be then substantial probability that a normal match between buyer and seller can be found. And the longer the seller can leave the property on the market, the higher the probability that a buyer will come along who really loves the property and is willing to pay an enormous price for it. (Of course, the price actually received for a property on the market a long time will depend also on the course of the market over that interval of time, and there is also a chance that the expected price will eventually decline with time as a property gets a reputation as damaged goods for having been on the market for too long a time.)

One concern in defining market values is that there may be, in effect, no market value because the property will probably not sell in what is conventionally regarded as a reasonable time. It is well known that in times of slack housing demand some houses may remain on the market for years, unsold. The above definition would suggest that the market price is the price that a seller would receive if the property *had* to be sold within a month, at whatever could be got for it. But that would not be a good indicator of market price for a potential buyer of that house, because in fact the buyer could not expect to be able to purchase at that price.

Consider a time when there are relatively few sales of houses because sellers are holding out for high prices, and buyers do not feel that they want to pay these prices. This situation is alleged to obtain in the down real estate markets we observe in many places around the world in the early 1990s, allegedly because sellers are accustomed to thinking that their properties have the values that they had recently, and are regretting not having sold earlier. If this is indeed the situation, there are in effect two prices for each house: the high price wanted by sellers and the low price offered by buyers.

Now of course, in any market period, there are *some* sales. What are we to make of these observed prices? Some have claimed that selling prices in such a situation are a measure of the high prices demanded by sellers, so that any index based on sales prices will be 'upwardly biassed'. This interpretation is suggested by the notion that sales observed are those made either to buyers who are urgent to buy, or to buyers who misperceive the value of a property and erroneously buy at the high seller's price. But another interpretation is equally plausible: that the sales observed are those made by sellers who are urgent to sell, or by sellers who misperceive the value of a property and erroneously sell at the low buyer's price.

There is no a priori reason to think that urgency to trade or errors in perceptions are more common among buyers or sellers, especially since most property-sellers are probably selling one property to buy another. If one is urgent to move to a different house, then one will have urgency both to buy one house and sell a different house. In this kind of slow market, then, the prices we observe will be both high prices paid by urgent buyers and low prices received by urgent sellers, and any index of prices of actual sales will be some sort of average of these low and high prices. Rather than have two price indices, one for prices for buyers in a reasonable time and another for sellers in a reasonable time, an index based on actual sales prices is essentially an average of these two indices.

A hedonic index or repeat-sales index based on actual sales prices might possibly take account of any 'bias' caused by a change in the average time taken to sell by including as a regressor the time on the market. Consider a slow market accompanied by price declines. Suppose that in this case many sellers become very patient in waiting for a good price, but buyers are unchanged by the recent downturn in prices. While it seems unlikely that there should be a marked difference in the patience of buyers and sellers, it is not inconceivable that there could be such a difference. Then there would be a tendency at these times for selling prices to reflect the values placed on properties by sellers rather than buyers, since the only properties sold are those that meet the seller's price. The time on the market would be a partial indicator of the proportion of sellers who are patient, and the coefficient of the time on the market in a hedonic regression might then be positive. We could then base an index on a standard house which was on the market for a standard time, by taking the fitted value in the regression for a standard house and a standard time on the market. The wisdom of doing this to generate an index for cash settlement of contracts, however, relies on some rather questionable assumptions. We might question the assumption that sellers are systematically more patient than buyers. If in the first part of our sample buyers were more patient than sellers, and in the second part of our sample sellers more patient than buyers, then time on the market by itself might get an insignificant coefficient in such a regression.

This problem could be fixed by including in the regression an interaction variable equal to the product of the time-on-the-market variable and time itself, but any such adjustments for time on the market are of questionable value, and time on the market may not be easily measured anyway. And there is still the question whether we want an index for cash settlement to measure the price of a house with standard time on the market; in the case where high average time on the market is a time of high seller patience, then buyers cannot expect to buy in a standard interval of time, and so the index does not represent the price at which they can expect to buy at the index-predicted price in that time interval.

Other models

There are other models in the statistical literature that might be applied to produce indices of prices of or cash flows accruing to claims on future incomes. Here, two index number construction methods will be considered, a factor-analytic method that treats quality measures associated with individual subjects as unobserved factors, and a selection bias correction method that is based on a model that predicts which subjects will be observed (which houses will sell). I have not tried implementing either of these models, and details remain to be worked out.

Factor-analytic models: interaction effects

It may be possible to produce indices of prices or rents on subjects of fixed quality, that is, to make adjustments for change in the mix of items observed, even if there are no hedonic variables that could indicate quality. The quality may be treated as an unobserved factor, and we may derive estimates of the factor and its time-varying loading into price, using methods analogous to those developed for factor analysis; see for example Lawley and Maxwell (1971). These estimated factors can be used to produce the indices that we seek. Note that this use of factor analysis is fundamentally different from that discussed in Chapter 5. There, factors represented shocks common to the values of many different claims on incomes for a given time period; here, the factors represent qualities common to many different time periods for a given subject.

Another way to view these methods is as interaction effect methods used in the analysis-of-variance. Note that our ordinary repeated-measures regressions are based on a model that is identical to the usual two-way analysis-of-variance model. Those who use these models commonly consider interaction effects. With an analysis of variance table with rows corresponding to individual properties and columns corresponding to dates, an interaction effect would mean that the quality of the property would have a different effect on price in different dates. However, since normally most of our cells have no observations and the cells normally have at most one observation, we cannot do the usual analysis-of-variance modelling of a general interaction effect. Still, a restricted interaction effect can be estimated, an effect that models the interaction effect as the product of a row effect and a column effect.

Intuitively, in the real estate example, if there is a period of time when there were a lot of houses sold that had low price, then we may feel that high-quality houses were under-represented then, and may want to give more weight to those high-price house sales we do observe in determining the index. Of course, we could not follow this procedure if there were only one observation per house: if there was a period when there were many low-priced houses sold we could not tell whether this was because the mix had shifted to smaller, less desirable homes or because there had been a drop in the level of house prices. But we expect to have multiple observations on each house; a house that is consistently low-priced relative to other houses may then be identified as a relatively less valuable house. Suppose we believe that price or rent follows the factor model:

$$p_{it} = \gamma_t + f_i + \lambda_t f_i + \varepsilon_{it} \tag{7.5}$$

where p_{it} is the price or rent of the ith subject at time t, γ_t and λ_t are parameters that vary through time and are the same for all subjects, f_i is the unobserved factor, the quality parameter that varies across subject but not through time, and ε_{it} is an error term that is independent of all other error terms, that is independent of $\varepsilon_{i't'}$ unless $i' = i$ and $t' = t$. Note that this is just a restricted form of the same model that gave rise to the hedonic repeated-measures regression in the last chapter, except that the hedonic variable is unobserved. The notation has been changed here to accord with that in the factor-analysis literature, but note that equation (7.5) is the same as (6.8) where f_i replaces δ_i, and s_{it} is constrained to equal f_i. This is a standard one-factor model from factor analysis; the only difference here is that here we do not observe p_{it} for all i for each t. What we have here is analogous to applying factor analysis to results of a pencil-and-paper exam where each subject (house) chose its own set of questions to answer (dates to sell).[8] The situation we are in is similar to the one that Wold (1966) studied when he sought to estimate the quality of horses from data on their performance in individual races; any given race excludes almost all horses.

As in all models in factor analysis, we must choose some normalization rule for the unknown factor. Note that the model would be unchanged if we divided all f_i by a constant and multiplied all λ_t by the same constant, and so these parameters are not completely identified. The usual procedure in factor analysis is to normalize the sum of the squared factors to 1.00; here we can say that the variance of the factor is 1.00.

This may also be thought of as a regression model that is nonlinear in the parameters, and unfortunately has a parameter to be estimated for each subject, possibly a very large number of parameters. Still, if we suppose that error terms for any given subject are uncorrelated with the error terms of all other subjects, then the variance matrix of the error vector ε has a simple form: it will be block-diagonal if the elements of ε are ordered properly. This means that the normal equations for the generalized least-squares estimate (the maximum-likelihood estimate if we suppose that the error vector is multivariate normal), an estimate that minimizes the quadratic form $\varepsilon'\Sigma\varepsilon$ where Σ is the inverse of the variance matrix of ε, have a simple structure. We can exploit the structure of the equations to produce an estimate of the coefficients without relying on general matrix inversion methods. If one differentiates $\varepsilon'\Sigma\varepsilon$ with respect to f_i and sets this derivative to zero, one derives fairly simple expressions. For example, if there are only two observations of subject i, and if the elements of the error term ε corresponding to this subject are independent and have the same variance, setting the derivative with respect to f_i to zero, and calling t_{i1} and t_{i2} the dates of first and second observations for subject i, we have:

$$\hat{f}_i = \frac{(p_{it_{i1}} - \hat{\gamma}_{t_{i1}})(1 + \hat{\lambda}_{t_{i1}}) + (p_{it_{i2}} - \hat{\gamma}_{t_{i2}})(1 + \hat{\lambda}_{t_{i2}})}{(1 + \hat{\lambda}_{t_{i1}})^2 + (1 + \hat{\lambda}_{t_{i2}})^2}. \quad (7.6)$$

Note that this is just a regression of price errors onto factor loadings for subject i (interpreting (7.5) for a given subject i as a simple regression model with one observation for each sale of property i, with $p_{it} - \gamma_t$ as the dependent variable, with f_i as the regression coefficient and, factoring out the f_i, with $1 + \lambda_t$ as the independent variable). Also note that the estimated quality factor $\hat{f}_i$ is chosen as a sort of compromise between the two observations on this subject; neither observation has a perfect fit in the regression equation. This means that, in inferring the level of prices on a particular date, the $\hat{f}_i$ on that date carry information about the quality of the ith subject that is partially independent of the information about that date. Estimates like these of the factors could then be substituted into (7.5) to allow estimation of the remaining parameters, i.e. the factor estimates could replace the s_{it} in (6.20). This suggests iterative procedures, like the iterative least-squares procedures of Wold (1966), to derive the generalized least-squares estimate without inverting large matrices. Estimation procedures for such models are discussed in Mandel (1961), for the case of complete data (every house sold every time period) and in Christoffersson (1970) for the incomplete data case, which is relevant here.

We could then take the estimated $\hat{\gamma}_t$ as a price index for a house of average quality. Doing this would enable us to control, in effect, for the changing mix of houses sold through time. If there were, for example, a period of time in which fairly few high-quality houses were sold and if the higher-quality houses were, let us say, dropping more in value at the end of the sample period, period T, than the lower-quality houses, then these observations might be fitted by a negative coefficient λ_T. A scarcity of observations of houses with high f_i might increase the standard error of λ_T, but may not induce any substantial bias in this coefficient.

We could also produce, from the estimated regression model, price indices for all levels of quality with which we are concerned. Since the normalization rule sets the standard deviation of the factor to 1.00, we can define an index, using $f_i = 1$, as the price index of houses one standard deviation above the mean in terms of quality. In the example in the preceding paragraph, an index of the price of a high-quality house would tend then to show more of a decline than an index of the price of an average quality house. There are, however, still problems to be worked out with these factor-analytic methods. Although Christoffersson showed that the maximum-likelihood method is consistent, he used a different normalization rule (that the sum of squared factor loadings is unitary). There are also problems of nonuniqueness and possible nonexistence of solutions to the equations defining the maximum likelihood solution.

Selection bias corrections

Selection bias models, that rely on estimated selection equations that describe what determines whether an individual subject finds its way into our sample, may be applied to the purpose of keeping our sights on an unchanging claim on future income in producing indices.

For a simple story about real estate prices, which might illustrate a motivation for modifying a method developed by Heckman (1976) so that it can apply to the construction of our hedonic repeated-measures indices, let us assume that house price at any date is linear in square feet of floor space (as in the examples in the preceding chapter) but that actual selling prices are contaminated by noise, let us say human errors. Sometimes buyers make the mistake of offering too much, sometimes too little. The assumption that the error term is pure human error and not an unobserved quality variable will motivate our hypothesis that the error term is uncorrelated with the independent variable, the square-feet variable, when there is no selection bias. Let us also suppose that, at certain dates, sellers of big houses are more willing to hold out, and refuse a sale if the buyer offers too little, than are sellers of small houses. At other dates, it may be sellers of small houses who are more willing to hold out. Supposing that we want our index to represent the expected price that one could expect to get for a house if one were a representative buyer and seller, both of whom would take the first offer, then clearly there is a time-varying bias in the coefficient of the square-feet-of-floor-space variable here.

A selection bias correction procedure could study which properties, based on their number of square feet of floor space, are less likely to sell, and to check whether those among these properties that actually did sell despite the low likelihood tended to sell for prices that are systematically higher or lower than predicted by the hedonic price equation. If so, we can infer something about the bias induced in the price equation from the under-representation of these properties in the sample.[9]

The procedure suggested by that used by Heckman (1976) is a two-step procedure, the first step involving the estimation of a probit model selection equation that explains which properties are observed.[10] The selection equation for each time period is a regression model $y_{it}^* = V_{it}\zeta + \varepsilon_{it}$, where V_{it} is a vector of characteristics of a property that affect its likelihood of sale, ζ is a vector of regression coefficients, and ε_{it} is an error term that is normally distributed with zero mean. In our example, V_{it} includes a constant term and the number of square feet of floor space. The probit model for time t asserts that property i is observed as sold at time t if and only if $y_{it}^* > 0$; we know whether a sale was observed but do not know y_{it}^*. In this model, as in all the usual probit models, the standard deviation of the error term ε_{it} is not identified; we could always double both ζ and this standard deviation without affecting likelihood, hence the standard deviation is normalized at 1.00. The parameter vector ζ can be estimated by the usual probit maximum-likelihood method using the data on characteristics of all properties and on whether they were sold.

In the second step, we estimate an equation with price as the dependent variable, with the hedonic independent variables and with additional variables that allow us to correct for the bias. This equation is supposed to describe prices of all properties, not just those whose price we observe, although we must base our estimation only on observed prices. For each property observed, we know from the fact that it was observed that the error term in the selection equation was greater than $-V_{it}\zeta$. If V_{it} is itself uncorrelated with the error term ε_{it}, then the distribution of the error term ε_{it} in the price equation conditional on both a sale and on V_{it} (and hence conditional on both a sale and the independent variables for property i at time t in the price equation) is truncated normal, truncated below $-V_{it}\zeta$. The conditional probability density of ε_{it} is then, above $-V_{it}\zeta$, the normal probability density

$\phi(\varepsilon_{it}) = (2\pi)^{-.5}\exp(-\varepsilon_{it}^2/2)$ divided by a factor so that the integral under this truncated curve equals 1.00. This factor is $(1 - \Phi(V_{it}\zeta))$ where $\Phi(\cdot)$ is the cumulative normal distribution function. The mean of this truncated distribution is the inverse Mills ratio:

$$M(-V_{it}\zeta) = \frac{\phi(-V_{it}\zeta)}{1 - \Phi(-V_{it}\zeta)} \,. \tag{7.7}$$

$M(\cdot)$ is a nonlinear function, monotonic, of positive slope, and concave upward. Moreover, suppose that the joint distribution of the error terms in the selection equation and the error term in the hedonic price equation is bivariate normal with zero mean, then the expected error in the price equation, given that the sale is observed and given $V_{it}\zeta$, is $\text{cov}(\varepsilon_{it}, u_{it})M(-V_{it}\zeta)$. Now assume that this covariance is constant through time and across all subjects; let us now add the estimated inverse Mills ratio $M(-V_{it}\hat{\zeta})$ in the hedonic price equation, as in (6.8), as an additional independent variable for each time period. We augment the matrix Z_A with an additional column for each time period, the tth such column is zero except for sales at time t, where it is $M(-V_{it}\hat{\zeta})$. When this regression is transformed, in the example of last chapter, by pre-multiplying by the matrix S, then we will have a modification of equation (6.20) with three more columns, one for each date (and three more coefficients corresponding to these columns). The tth additional column consists of zeros unless the first sale occurred on that date, in which case the element is $-M(-V_{it}\hat{\zeta})$, or unless the second sale occurred on that date, in which case the element is $M(-V_{it}\hat{\zeta})$. (We would not normally expect to find that these three columns sum to zero, but the first of these three columns might be dropped if there is a problem.) The second step of this procedure is to run this regression obtaining estimates of the coefficients. Then $\hat{\gamma}$ is a consistent estimate of γ in the sense of increasing the number of observations without increasing the number of parameters in the model $y = z\gamma + \varepsilon$. After the price equation is estimated, the index may be produced in any of the ways described above from the estimated $\hat{\gamma}$.

Since in this example we included in the selection equation as independent variables only the independent variables in the price equation (constant and square feet), the $M(-V_{it}\hat{\zeta})$ would be a non-linear function of the other included variables. Since the assumption of linearity in the other included variables in the price equation is not based on any real knowledge, and is just a working hypothesis, our ability to identify any selection bias through this procedure may be limited. The assumption of linearity in the selection equation may be equally problematic.

Interpretation

For index number construction, there are many possible approaches, many models, many alternative sets of prior restrictions, many alternative sets of hedonic variables. Choices that must be made involve human judgment. Human judgment, in deciding which method to use, plays every bit as large a role in the specification of real estate or national income indices for settlement of contracts as it does in the tasting room at the coffee exchanges, where the quality of coffee is assessed. Fortunately, with the quantitative data and array of methods at our disposal, we can decide upon methods of index number construction whose later implementation may require little or no judgment, so that the definition of the index becomes objective.

The problem of change in mix of subjects observed has been a major issue in this and the preceding chapter. We have seen three approaches for dealing with this problem. The first approach, stressed in the preceding chapter, involved the use of subject dummies in a hedonic regression; the inclusion of these dummies in the regression meant that index changes were inferred only from changes in the price (or income, if that were used to form the dependent variable) of individual subjects. This first method handles the effects of unobserved quality by assuming that the effect of the unobserved quality variables on price is constant through time. We could not put a dummy for each subject-time period, to control both for subject effects and time effects, because that would exhaust all degrees of freedom. The second method, based on the factor model, equation (7.5), allows the effects of quality variables to vary through time, but requires that the time variation follow the same pattern for all subjects, thereby conserving degrees of freedom and allowing estimation to take account of the effects of unobserved quality. This method results in a modification of the hedonic repeated-measures regressions (as exemplified by (6.20) or (6.26)) by substituting quality $\hat{f}_i$, estimated as an unobserved factor that is constant through time, in place of the hedonic variable s_{it}. The third method, the selection bias correction method, relies on an assumed lack of independence of residuals in the hedonic equation from variables in the selection equation, and on functional form assumptions for the distribution of error terms. This method results in a modification of the hedonic repeated-measures regressions (as exemplified by (6.20) or (6.26)) by substituting the estimated inverse Mills ratio $M(-V_{it}\hat{\zeta})$, estimated using a selection equation, in place of the hedonic variable s_{it}.

There are certainly other methods one could use to correct for change in mix; choices among methods must depend on the plausibility of assumptions. Unfortunately, there is no final arbiter that will tell us whether we have made the right choices, except for the market. We do not find out, even years later, whether our index number values were really right or not. But if people want to hedge in the market that is created, then the index is a success.

8
The Problem of Index Revisions

Most published economic indices are revised after they are first published. Information does not come in all at once, and timely publication dictates that the preliminary index numbers be later revised. The repeated-measures indices developed in the preceding chapters are vulnerable to revisions after especially long intervals of time. They have the property that, unless the repeated measures come sequentially, i.e. every period, there will be revisions in the indices after the index numbers are first produced, even if the raw data used then were perfectly accurate and complete. Repeated measures may not come sequentially. In constructing real estate price indices, we obtain a measure only when the real estate is sold, which is very infrequently. In constructing income indices, such as national income indices, that are based on repeated measures of individual incomes, cost and good-will considerations would discourage us from measuring the same individuals every time the index is produced. It may not even be possible to get an unbiased sample measured every time period.

There are other index number construction methods, such as ordinary-least-squares regression-per-period hedonic regressions, that do not normally produce revisions. There would seem to be an advantage to index number construction methods that do not produce subsequent revisions, if the index numbers are to be used to settle contracts. It will certainly be troublesome to those who lost money in the settlement of contracts to discover later that if the settlement had been based on the revised value of the index, they would not have lost money. How should we handle such revisions, then? Should we confine ourselves to index number construction methods that do not produce revisions? Whether revisions are routinely produced should not be an issue; the issue instead should be to choose the index number construction method in consideration of the effect of initial index errors on hedging risks, given that settlements will not be revised.

The problem of revisions is a fundamental one. The revisions reflect genuine information that arrives later, after the date to which the index number applies. That an index number construction method does not produce revisions is not a virtue if new information tends to arrive implying revisions and this information is just being ignored.

To understand the potential for subsequent revisions in repeated-measures indices, let us consider the prospects for revisions in the real estate price indices of the simple, ordinary repeated-measures form, the form of the original repeat-sales regression of Bailey *et al.* (1963). This index is produced by running a regression using independent variable matrix z and dependent variable vector y, as exemplified by expression (6.10). Consider how this index number construction method infers price changes between time period $t - 1$ and time period t. (For an explicit example, subtract (6.16) from (6.17).) Part of the information used to infer this price change is data on price changes of those properties that sold both at $t - 1$ and also at t. But presumably there are very few such properties. This method, if used at time t to produce the index change $\hat{\gamma}_t - \hat{\gamma}_{t-1}$, also makes use of data on price changes of groupings of properties sold twice; for example, a pair of properties sold first at $t - 1 - k$, $k > 0$, one sold for the second time at $t - 1$ and one sold for the second time at t. Each such pair of properties gives an indication of the change in housing prices between $t - 1$ and t, and the repeat-sales regression method makes optimal use of all of these pairs in inferring the index. With data on current and historical sales available at time t, we expect to find many more such groupings of properties than we can expect to find single properties that sold both at $t - 1$ and t, and hence the repeat-sales regression estimates tend to be dominated by the information in such groupings of properties, rather than in the individual properties sold both at $t - 1$ and t.

Now, in considering the potential importance of subsequent revisions in the index between $t - 1$ and t, consider that after time t there will be other groupings of properties sold twice that can also be used to infer price changes between $t - 1$ and t: for example, a pair of properties sold the second time at $t + k$, $k > 0$, one sold for the first time at $t - 1$ and one sold for the first time at t. Symmetry suggests that there will be as many of these kinds of new informative groupings of properties, observed after time t, as there were such informative pairs of properties observed at time t.[1] This means that, from these symmetrical observations alone, we can expect our data set, relevant to estimating the price change between $t-1$ and t, effectively to double eventually; not right away, but when properties sold for the first time at $t-1$ and at t sell again; this could be years, even decades, later. Moreover, there will also tend to be even more information than this symmetry argument suggests. For estimating price changes between $t-1$ and t, we will, after time t, obtain data from both strictly before time t and strictly after time t that carries information about the price change between $t-1$ and t. For example, we may find a trio of properties, the first sold at time $t-2$ and again at time $t+1$, the second sold at time $t-2$ and $t-1$, and the third sold at time t and time $t+1$. Thus, ultimately, we expect to see more than a doubling of relevant information about the price change between $t-1$ and t.

Note that the same symmetry argument that was used to infer the importance of subsequent revisions in the index also can be used to infer the importance of these revisions to individual hedgers: for every hedger who was hedging a property purchased at time $t-k-1$ and sold at time t there will be, by symmetry, a hedger who is hedging a property purchased at time $t-1$ and sold at time $t+k$. The latter hedgers, who will be receiving a settlement from their futures position at time t, do not yet know at time t the second sale price of their property; the revisions are related to such second-sales prices.

This intuitive argument based on symmetry suggests, then, that, so long as the variance matrix for errors in the regression model treats past and future symmetrically, we would expect the standard errors of the estimated one-period index change to decline, over years and decades after time t, eventually by a multiplicative factor that is less than one over the square root of two. Longer-period index changes may tend to show less of an improvement in standard errors ultimately, since longer-period changes are sums of one-period changes, the earlier values of which are already substantially revised when the longer-period index change is first reported. How important this reduction in standard errors is depends, ultimately, in the context of the regression model, on how small the standard errors at time t are. If the standard errors are very small, then we may feel that we needn't be bothered by revisions. In the context of a real estate price index, we can obtain small standard errors by making the index cover a wide geographical area, so that there are very many observations of sales. But those who want to use the index to settle contracts may not always want a very large geographical area to be used to compute the index. Moreover, we can never be sure, just from looking at the standard errors produced for the index at time t, that the subsequent data will be generated from the same statistical model, and so we can never be sure that the revisions will be as small as would be suggested by looking at the estimated regression standard errors.

It should be noted, too, that revisions do not disappear completely even after a time period greater than the greatest interval between repeated measures. Suppose, for example, that we are constructing a repeated-measures index of national income by sampling some individuals every period and, to save costs, sampling other individuals every other period. Suppose that a third of our sample of people is questioned every period about their income, a third is questioned only in even-numbered periods, and a third is questioned only in odd-numbered periods. No individual is left unmeasured for more than one period. An ordinary repeated-measures index will be found continually to revise (if only slightly) the entire history of the index every time a new time period is added. Since all values of the index are linked together by these repeated measures, they are all revised when any of the index values is revised.

Variance components in regression-per-period hedonics

The regression-per-period hedonic index number construction method described in Chapter 6 is done in such a way that each index value is produced independently of the others and there are no subsequent revisions of the index. But the model that gave rise to this index number construction method was deficient in not recognizing a dependency across time periods of regression error terms, due to the similarity through time of the errors for individual subjects. When we stack the regressions of the regression per period methods into a giant matrix, as in expression (6.5), then we should allow, if we are not including subject dummies, a variance component to the error term, a component that is associated with the individual subject. Certainly, individual subjects tend to have idiosyncratic features that are not described by independent variables, and there should be some tendency for these features to persist through time.

Assuming there are such variance components to regression error terms, ordinary least squares (which amounts to a separate regression each period) would be unbiased, under the usual assumptions of the regression model. But the ordinary-least squares regression estimates will be inefficient. Taking account of the variance component of the error term could dramatically improve the efficiency of estimation. To see this, suppose, in the real estate example, that there was very little noise in individual sales prices, that the error terms induced by the selling process is small—most of the error in the regressions are due to unobserved quality differences across properties. A generalized least-squares estimate of the parameters in the model defined in terms of (6.5), with a variance matrix with the structure indicated by the variance components model, would in effect use the information about repeat sales to improve the efficiency of estimation, without dropping the information in the single-sales observations altogether. The efficiency gain in this case would be dramatic, since the variance components model estimator can take account of the fact that price changes for individual properties have virtually no error. A consequence of this efficiency gain would be that we will find (small) revisions in past values of the index whenever another time period goes by that adds another data point to the time series of the index; but these revisions could be dwarfed by the overall reduction in standard errors made possible by using the variance components.[2] It should be clear from this example that, in principle, the absence of revisions in the regression-per-period hedonic regressions is by no means an advantage for these methods.

Interval-linked indices

Publishers of price indices, whether they are used to cash settle contracts or not, customarily publish only one index number at a time, and do not publish revisions each period of the entire history of the index. If the index is monthly, then we see a single new number each month, not a revision each month of the entire index for, let us say, hundreds of months. There is a good reason why only a single number is published: it is costly to publish so many numbers, and costly for the public to assimilate and use the revisions each period. Part of the costs to the public of using such frequent revisions is that there will be confusion and errors, errors such as combining numbers in an analysis from more than one revision of the time series. If the revisions are not too important, then it is not worth these costs to publish such revisions.

Suppose, then, that we operate under the constraint that we can never revise past values of the index: each time period a new index value is published, and this new value is appended to the time series of past published values of the index. The most obvious way to proceed would be to use one of the methods described above to produce the entire index each period, including all revisions in the past values of the index, and then simply throw away all of the revisions of the past values of the index, publishing only the latest value. But this is not generally the right way to proceed, since the latest value was in effect produced under the assumption that all past values would be revised. The new data may, for example, have occasioned similar upward revisions in the index for all of the months of the past few years. If we then publish an index value for the current month, expecting users of the index to use this new number along with past published numbers of the index, then users of the index would incorrectly infer that there was a sudden jump in prices in the last month alone.

What is the optimal thing to do if we are publishing only one number each month? It is critical, in answering this question, to clarify to what purpose the index will be put. Here, the primary purpose of the index will be to cash settle contracts, such as futures contracts. Consider, then, what the revisions really mean to persons who wish to trade in conventional short-term futures contracts cash-settled on the basis of our repeated-measures price index.

If there were no revisions in past values of the index, changes measured over any interval ending at time t would be perfectly correlated (conditional on information available at time $t - 1$) with the level index at time t, and so a one-period futures contract settled in terms of the level of the index is as well suited to hedgers of any given horizon as it is to hedgers of any other horizon. When, however, new information arises that results in revisions in past values of the index, then we have the problem that there are really produced each period an array of best estimates for changes over all past intervals of time. If we arbitrarily chose to settle a futures contract on the basis of the level of the latest estimated index, then we would be choosing to settle on the basis of the best possible estimate at the time of settlement of the change since the base period. If we kept the base period fixed through time, then, as time progresses, the futures contract would be being settled on changes over longer and longer intervals of time. This situation could cause serious problems for a short-term hedger: between the time he or she took out the hedge and the time the position was closed out, there could be new information about price changes *before* this person took a position in the futures market, and there would be a cash settlement even if there was no change in cash-market price while this hedger held a position in the futures market. Obviously, one should not design a contract to settle based on such an arbitrary formulation.

This means that settlement must be based on an estimate of some consistently defined change in price as of the time of settlement. Of course, making the settlement in terms of a change in price over a fixed interval does not mean that we must quote the settlement in terms of a change in an index. We can still base the settlement on the level, rather than change, of an index, so long as the index is what will be called here an interval-linked index.

The simplest interval-linked index is one in which the level of the index at time t is computed as the level at time $t - k$ as published at time $t - k$ plus the best estimate at time t of the change in price between $t - k$ and t. Since the market knows the published value of the index at time $t - k$, all of the uncertainty about settlement concerns uncertainty about the best estimate as of time t of the change between $t - k$ and t.

In designing indices for the settlement of contracts, it is important to consider a more general form for the interval-linked indices, which will be sorts of averages of the fixed-interval indices over an array of intervals, the array chosen to reflect the preferences of users of the index.

The use of an interval-linked index may benefit those who sign contracts settled in terms of the index, in that the interval-linked index may be so devised as to compromise among their various interests. It will also benefit others who wish to use an index, such as people signing labor contracts which have escalator clauses tied to the price of real estate in their area. They too, like the users of futures contracts, are interested in price changes over certain intervals, and an index might be chosen to reflect their concerns. The interval-linked index may not be ideal for those who engage in forward contracts or swaps, and who are bypassing the liquid futures markets in order to have their special needs addressed. For these, who are signing a contract that has the horizon that they are interested in, it may be better to settle the contract on a price change over an interval of particular interest to them.

It is, of course, conceivable that there could be many futures contracts; each futures contract could be settled on the basis of the best estimate on the settlement date of price changes over a specified interval of time, where the interval is unique to that contract. This means that an estimated price change must be created specifically for each contract, and that different contracts cannot be settled based on the same index. Opposing this, however, are considerations of liquidity and public information. We cannot have too many futures contracts; the number of contracts that can be traded is necessarily limited: the public cannot readily process the information in estimated price changes over many different time intervals. It is natural, then, to consider settling contracts on the basis of an index that reflects some sort of weighted average of best estimates of price changes over various intervals of time, where the weight on each interval reflects the concern in the market for price changes over that interval.

The interval linked index I_t is defined in terms of its own lagged values and estimated changes in prices by the formula:

$$I_t = \sum_{\tau=0}^{\infty} a_{\tau t}\left(E_t(\bar{P}_t - \bar{P}_{t-\tau-1}) + I_{t-\tau-1}\right) \tag{8.1}$$

where

$$a_{\tau t} \geq 0 \ , \quad \sum_{\tau=0}^{\infty} a_{\tau t} = 1 \ . \tag{8.2}$$

Here, $\bar{P}_t$ is the aggregate price level at time t: if $\bar{P}_t$ is measured in logs, then this is a geometric index, if in levels rather than logs, an arithmetic index. Each coefficient $a_{\tau t}$ is the weight given to expected price changes over the interval of length τ in determining the index at time t. The reason for choosing an interval-linked index is so that the uncertainty at time $t - 1$ about the index at time t is distributed over a number of intervals of price change, not just the price change over a single time interval.

Since all past values of the index are known, there is no uncertainty at time $t - 1$ about any of these terms in (8.1). Technically speaking, as far as futures trading goes, we might as well substitute *any* numbers for the lagged index values, so long as these numbers are announced to the public each period. In practice, it makes much more sense to use the lagged published values of the index numbers, since these have already been published, and since they are estimates of the lagged price level.

There is also the question of 'starting' the index. If we followed (8.1) from the very beginning of our historical sample in time 0, then the index for period 1 would be determined only by properties sold in period 0 and 1, probably a very small sample. But we do not need to follow (8.1) from the very beginning of our sample; only from the beginning of contract settlement. We might then use one of the simultaneous index number-generating formulas, say the IGRS or IVWARS described above, to arrive at historical values for the index before the trading begins. Then the index in period 1 would be based on the entire sample of data before the beginning of futures trading.

Let us now derive the optimal weights a_τ, $\tau = 0, 1, \ldots$ for the interval-linked index with conventional (not perpetual) futures markets. For this purpose, the t subscript in $a_{\tau t}$ may be dropped, since the structure of the optimization problem dictates that the weights will be constant through time. This derivation is intended to be largely illustrative, since we cannot be sure of some of the assumptions that underlie this derivation. Our maintained hypothesis is that there is a common factor $\bar{P}_t$ for the prices, so that the price of the ith investment property equals $s_i\bar{P}_t$ where s_i is a factor loading reflecting the size and quality of the property. We disregard here the price variation in individual houses that is idiosyncratic to that house, since the variation is not hedgable. In factor models, there is of necessity a normalization rule for the factor, otherwise we could double the factor and cut in half all s_i values; we shall suppose that the factor is normalized to a fixed number in the base year.

Consider a hedger who at time t buys the ith property, planning to sell it at time $t + n$, $n > 0$. Suppose that the hedger sells s_i contracts in the futures market, to hedge this risk. We are here supposing that the hedger knows s_i; of course we do not assume that the constructor of the index number knows s_i for each property; this is the reason for estimating repeat-sales price indices described above. Let us suppose that the contracts are all one-period contracts, that there is only one settlement, the final cash settlement at the end of the period, and that the hedger rolls them over, always selling s_i contracts in each period. It would generally not be optimal to sell s_i contracts—the number of contracts should be adjusted each period to reflect the accuracy of the hedge—but we will assume a fixed number of contracts.

The settlement of each contract as it matures will reflect the new information that arises over the preceding time period. For expositional purposes, let us make assumptions so that the settlement at time t is nothing other than the innovation in the index I_t. (This is assuming there is no backwardation or other expected changes in the futures price.) This means that the settlement for a single contract is:

$$\Delta_t I_t = \sum_{\tau=0}^{\infty} a_\tau \sum_{j=0}^{\tau} \Delta_t e_{t-j} = \sum_{\tau=0}^{\infty} b_\tau \Delta_t e_{t-\tau} \tag{8.3}$$

where $e_t = P_t - P_{t-1}$ and $b_\tau = \Sigma_{i=\tau}^{\infty} a_i$. The symbol Δ_t is used here to denote the innovation operator $E_t - E_{t-1}$. (If $\bar{P}_t$ is measured in logs, $\Delta_t I_t$ is not the dollar settlement but the log of the final futures price minus the log of the initial futures price.) The unhedged risk of the ith investment property would then be given by:

$$u_{it} = s_i \left(\sum_{k=1}^{n} e_{t+k} - \sum_{k=1}^{n} \sum_{\tau=0}^{\infty} b_\tau \Delta_{t+k} e_{t+k-\tau} \right). \tag{8.4}$$

If $\bar{P}_t$ is measured in levels, then this unhedged risk is the change in price of property i plus the (offsetting) change in price of the (short) futures contracts. If $\bar{P}_t$ is in logs, then this is the log of the proportional change in the value of property i divided by the proportional change in the futures price. We want now to find the coefficients b_τ, $\tau = 0, 1, \ldots$ that minimize the variance of this unhedged risk, thereby making the best possible contract for the hedger. Of course, we could design a contract particularly for this hedger, so that settlement is based on the price change between the dates that concern this hedger. But we cannot expect to make liquid markets in all such time intervals. By constraining the problem to that of minimizing the variance of the unhedged risk with respect to the values of b_τ $\tau = 0, \ldots,$ we thereby treat persons with all start dates and all end dates of the interval symmetrically, all of them hedging in the same market.

We do not yet have the problem specified well enough to derive the optimal values of b_τ, $\tau = 0, 1, \ldots$. We need next to know something about the stochastic properties of the process e_t. Let us suppose that e_t is white noise (serially uncorrelated), i.e., that $\bar{P}_t$ is a random walk. If $E_t\bar{P}_t$ is derived from a repeat-sales estimator (interpreting the regression coefficients in a Bayesian manner from uninformative priors, as posterior means conditional on information), then clearly there will be no information about e_t until time t. This means that innovations in e_t occur strictly *after* time t, and that innovations $\Delta_T e_t$ are zero if $T < t$. This is just the opposite of the situation that is usually assumed in economics, where information about an economic variable e_t is gradually uncovered until the date t, at which time the true value of that e_t eventually becomes known, and uncertainty is then zero. In our present application, nothing becomes known until time t because we do not learn about prices until we see sales. If we had adopted a price estimator that made use of information about the distribution of price changes—say the standard deviation of price changes or the serial correlation of price changes—then we would of course start to acquire information about price at time t before time t. For the sake of argument now, or for the reasons of objectivity outlined in the section on Bayesian methods in the preceding chapter, let us suppose that no such information is used to arrive at the index at time t, and indeed nothing is known about prices at time t until time t. Let us also assume (for expositional neatness) that the variable e_t eventually becomes known, at $t = \infty$. These assumptions imply that e_t is the sum from $j = 0$ to ∞ of its innovations $\Delta_{t+j}e_t$. We still need to know more about the stochastic properties of e_t. Let us assume that $\Delta_{t+k}e_t$ is uncorrelated with $\Delta_{t+k}e_{t+j}$ for all j not zero, and that the variance of $\Delta_t e_{t-\tau}$ equals σ_τ^2, independent of t. Under the assumptions of the three-period case (three time periods,

0, 1 and 2), as in the example in equation (6.5) displayed above, this lack of correlation is assured regardless of the number of observations of houses sold for each interval between sales. (However, this conclusion depends on the assumption of homoskedasticity of the error terms in the regression equation, and, moreover, complete lack of correlation for all j and k does not extend to the case of four or more time periods without restrictions on the number of observations of houses sold for each interval between sales.)

The variance v_{in} of the unhedged risk, then, for hedger (i) who plans to hold a property for n periods, rolling over one-period futures contracts for n consecutive periods to hedge the risk of this property, is:

$$v_{in} = s_i^2 \left(\sum_{\tau=0}^{n-1} \{(n - \tau)(1 - b_\tau)^2 + \tau b_\tau^2 + \tau\}\sigma_\tau^2 + n \sum_{\tau=n}^{\infty} (b_\tau^2 + 1)\sigma_\tau^2 \right). \tag{8.5}$$

Let us now derive the optimal coefficients b_τ, $\tau = 0, 1, \ldots$. It is clear that $b_\tau = 0$ for $\tau > n - 1$; there is no point in allowing the settlement today to be affected by information about price changes that occurred before any of the currently active traders were in the market. Differentiating with respect to each of the b_τ, $\tau = 0, \ldots, \infty$ and setting to zero, we find:

$$b_\tau = \frac{n - \tau}{n}, \quad a_\tau = \frac{1}{n}, \quad \tau = 0, \ldots, n - 1 \tag{8.6}$$

and $b_\tau = 0$ for $\tau > n - 1$. Note that the coefficients do not depend on the variances of the innovations σ_τ^2. Of course, the importance to the behavior of the index of the weights b_τ depends on the variance of the innovations: if the bulk of the information about e_t is quickly revealed, then there is not much effect of b_τ for large τ on the properties of the index. Note that, even though all hedgers are presumed to plan to hold (and hedge) their property, when they purchase it, for n periods, the optimal interval-linked index shows that estimated price changes over all intervals from 1 to n are given weight in determining the index, not just the estimated n-period price change. The reason is that while each hedger starts out with an n-period horizon, the horizon gradually shrinks as time approaches the sale date.

Now, of course, those who hedge in futures markets do not generally know the exact date of the ultimate sale of the property hedged. It will be better to suppose that the number of periods n to the second sale date has a probability distribution $f(n)$, a distribution reflecting the pattern of actual sales. What then is the optimal set of weights b_τ, $\tau = 0, ...$? There is no easy answer to this question; ideally we would embed the problem in the intertemporal life-cycle maximization problem that the investor-hedger faces, but any such well-delineated problem will yield no easy solution. We can, however, ask what set of weights b_τ, $\tau = 0, ...$ minimizes a probability-weighted present value v of variances v_{in}:

$$v = \sum_{n=1}^{\infty} \rho^n f(n) v_{in} . \tag{8.7}$$

Substituting equation (8.5) into the above expression for v, differentiating with respect to b_τ, $\tau = 0, ...,$ setting the derivatives to zero, and solving, one finds that the coefficients b_τ are given by:

$$b_\tau = \frac{\sum_{n=\tau+1}^{\infty} \rho^n f(n)(n - \tau)}{\sum_{n=1}^{\infty} \rho^n f(n) n} . \tag{8.8}$$

Note that b_0 is always one, and so the sum of the a_i is always one. Note also that as before the weights b_τ and a_τ are unaffected by the variances σ_τ^2.

An important special case is that in which the probability distribution for the interval n between adjacent sales is exponential, proportional to p^n. This distribution accords with the binomial distribution for number of sales discussed in the preceding chapter. Substituting into the above formula, and calling $\theta = \rho p$, we find then that $b_\tau = \theta^\tau$, so that the weights $a_\tau = (1 - \theta)\theta^\tau$; the weights a_τ also follow the exponential distribution. This means that, if there were no discounting ($\rho = 1$), the optimal a_τ are the same as the proportion of sales of interval τ.

There are a number of difficult issues that need resolution before one can feel assured that one has found the optimal interval-linked index. Individuals' hedging needs cannot be considered in isolation; the specification of an optimal interval-linked index should ultimately be made in terms of the entire maximization problem that the individual faces. One aspect of this global maximization problem is that individuals who sell a property (house) may plan to purchase another asset in the same market. Thus, the raw turnover rates of sales are not good indications of the probabilities that are relevant to specifying the optimal interval-linked index; rather, we need to know the probabilities that people will shift from one real estate market to another or out of the market altogether. Moreover, we need to know the correlation of prices in one market with the market the persons tend to switch to. Another issue that merits consideration is the behavior of pure speculators, rather than hedgers. One might suspect that speculators would prefer to have contracts settled on indices that link from rather recent dates.

We do not now have the information that would permit the rigorous specification of the optimal interval-linked index. Moreover, the weights that are derived as optimal will change from time to time, as investor constraints and trading patterns change. It would be undesirable to change the method of index number construction too frequently, doing so compromises the acceptance and understanding of the market. But certainly *some* form of interval-linked index is appropriate for cash settlement of futures contracts. The exact form of the interval-linked index may not matter too much. So long as revision errors are not too large, the different interval-linked indices may behave relatively similarly.

Indices that are derived by conditioning on lagged index values

Alternatives to the interval-linked indices are indices based on the same regression approaches that were discussed above, but conditioning on *all* past values of the index; that is, treating these as known before estimation. Estimation proceeds under the assumption that the past information that will be available to users of the index is correct, the same assumption that users are making. It is easy to modify regression model estimates to condition on certain coefficients. In the regression model one merely modifies the dependent variable and omits all independent variables that correspond to coefficients whose values are constrained. The modification of the dependent variable is achieved by subtracting, for each observation, the constrained value for each constrained coefficient times the observation of the corresponding independent variable. In other words, one brings over these independent variables times their coefficient values onto the left-hand side (the dependent-variable side) of the regression equation, and estimates the remaining coefficients as usual.

Consider, then, the modification of the IGRS (WRS) index described in Chapter 6 to constrain it so that all past values are given. We may call this the C-IGRS index, for constrained IGRS index:

$$I_{\text{C-IGRS}t} = \sum_{i \in s_t} \left(w_{it} (p_{it} - p_{it_{it}} + I_{\text{C-IGRS}t_{it}}) \right) / \sum_{i \in s_t} w_{it} \tag{8.9}$$

where s_t is the set of all i (of all investment properties) for which there is a sale, not for the first time for that property in our data, at time t, where w_{it} is the weight given by the IGRS index to the repeat-sales pair (pair of sales of a single property) for which the second sale corresponds to property i at time t, and where t_{it} is the date of the sale of property i in the sale preceding that at time t. Note that this is just the general form of expression (6.19) above, the last normal equation of the IGRS normal equations. The last normal equation is the derivative of the sum of squared residuals with respect to the last (latest) index value, set to zero; we are minimizing the weighted sum of squared residuals with respect to only the latest index value.

There is also an analogy to (8.1) above. Suppose that the weights w_{it} in (8.9) depend only on the interval between sales (as with the heteroskedasticity corrected indices proposed by Case and Shiller (1987), Goetzmann (1990), and Webb (1988)); then if we define $a_{\tau t}$ as the sum over i of all weights w_{it} in (8.9) for which the property i, selling in time t, also sold in time $t - \tau - 1$, divided by the Σw_{it}, then (8.9) is the same as (8.1) where $E(\bar{P}_t - \bar{P}_{t-\tau-1})$ is taken to be the average log price change of all properties sold in time t, $\tau + 1$ periods after the previous sale.

Of course, this average, the average log price change of all properties sold in time t, $\tau + 1$ periods after the previous sale, is not a very good measure of the aggregate price change between $t - \tau - 1$ and t. There might be very few properties that sold on both these dates, and hence the estimated change derived from these properties alone will be very inaccurate when compared with an estimated price change derived from one of the regression models. But we shall see that, if weights w_{it} depend only on the intervals between sales, (8.9) is the same as (8.1) where $E(\bar{P}_t - \bar{P}_{t-\tau-1})$ for all τ are replaced by the corresponding IGRS estimated price changes with the data set ending in time t. Assuming that w_{it} depends only on the interval between sales, (8.9) implies, using the $a_{\tau t}$ described just above, that:

$$\sum_{\tau=0}^{\infty} a_{\tau t}(I_{\text{C-IGRS}t} - I_{\text{C-IGRS}t-\tau-1}) = \sum_{\tau=0}^{\infty} a_{\tau t} \operatorname*{mean}_{i \in s_{t\tau}}(p_{it} - p_{it_{it}}) \tag{8.10}$$

where $s_{t\tau}$ is the set of all repeat-sales pairs such that the second sale was in time t and the first sale was $\tau + 1$ periods earlier. Replacing the γ_τ estimated using the IGRS estimator with data through t with $E_t(\bar{P}_\tau)$, $\tau < t$, in the normal equations of the IGRS estimator (as exemplified by (6.18) and (6.19)), it follows that:

$$\sum_{\tau=0}^{\infty} a_{\tau t} E_t(\bar{P}_t - \bar{P}_{t-\tau-1}) = \sum_{\tau=0}^{\infty} a_{\tau t} \operatorname*{mean}_{i \in s_{t\tau}}(p_{it} - p_{it_{it}}) . \tag{8.11}$$

Since the right-hand sides of the above two equations are the same, we may set the left-hand sides as equal to each other. Rearranging the resulting equation, one finds that C-IGRS satisfies (8.1) using IGRS estimates in place of the $E_t(\bar{P}_\tau)$ and is in this sense an interval-linked index.

Note that under the assumptions of the regression model that gave rise to the Bailey–Muth–Nourse estimator discussed above, the Gauss–Markov Theorem implies that the best linear unbiased estimate of any linear combination of the price indices is the corresponding linear combination of the estimated regression coefficients. When we base our index instead only on the last normal equation (corresponding to the equation zero equals the derivative of the sum of squared residuals with respect to the latest index value) to determine the latest index value taking earlier index values as given, then only one linear combination of the regression coefficients is optimally estimated; the linear combination using the weights $a_{\tau t}$ defined in terms of the w_{it} as above.

In the case where the distribution of waiting times between sales follows the exponential distribution as described above, so that the equation $a_\tau = (1 - \theta)\theta^\tau$ gives, using (8.8), the optimal weights, and when there is no discounting, the optimal weights a_τ are the proportion of sales of interval τ between sales. Using a heteroskedasticity correction that down-weights sales over longer intervals is the same as putting in some form of discounting of future error variances.

The simple index number formula (8.9) therefore has two different interpretations. The first interpretation is as a conditional estimate of the price level, conditional on all past values of the index. The second interpretation is as an interval-linked index of the kind described in the preceding section, where the weights given to various intervals between sales reflect the actual frequency of trading at the intervals.

A constrained interval value-weighted arithmetic repeat-sales index (C-IVWARS), analogous to the IVWARS index of above, is given by:

$$I_{\text{C-IVWARS}t} = \frac{\sum_{i \in s_t} w_{it} P_{it}}{\sum_{i \in s_t} w_{it} P_{it_{it}} / \text{index}_{\text{C-IVWARS}t_{it}}} \tag{8.12}$$

where s_t, t_{it}, and w_{it} are defined as for equation (8.9) above. This is the last normal equation of an instrumental-variables estimator $(z'\Sigma_{11}x)^{-1}z'\Sigma_{11}y$, equation (6.25). The last normal equation there is, of course, just the requirement that the residuals be orthogonal to the last instrument; the last instrument is a vector which is zero everywhere except for observations where there was a second sale in the last time period, where the element of the vector equals one. Equation (8.12) asserts just that the weighted average of errors (in predicting the second sale price using the index and the first sale price) is zero.

Interpretation

Before we can make an informed decision about index construction methods, we must collect information about the hedging needs of market participants. Moreover, if, as time goes by, the hedging needs of these people change, then the optimal index number construction method must also be changed.

Clearly, much work remains to be done on the optimal specification of index number formulae. Still, the basic idea of a distributed interval-linked index presented here would seem to be applicable to many situations; certainly we are unlikely to want to choose an arbitrary single price change interval to settle contracts.

Index number formulae that were derived by conditioning on lagged values were not derived as optimizing a hedger's problem. But these indices, recommended by their simplicity, have, we have seen, some resemblance to interval-linked indices, and may be useful for settling contracts.

There has been no discussion in this chapter of the possibility that contract settlements might be revised later, when the index on which settlement was based is later revised. The minimization problem that derived the optimal interval-linked index for settlement purposes did not allow lagged settlements or revisions of settlements. Generalizations of the theoretical framework presented here could be used to investigate such alternative settlement procedures.

There was also no discussion in this chapter of the optimal construction of repeated-measures rent or income indices to serve as the basis of cash settlement of perpetual futures contracts. The minimization problem described above might also be altered to consider specification of the optimal rent or income index.

For some contracts, such as real estate, it is probably possible largely to finesse the revision problem by merely choosing a geographical area large enough to ensure that standard errors in the indices are very low, so that revisions are not important. Fortunately, it appears that there are substantial comovements of real estate, not only over entire metropolitan areas, but also over megalopoli or substantial geographical regions, so that large geographic areas indices might be used to minimize the revision problem for settlement purposes.

9

Making It Happen

I have attempted in this book to carry to its logical conclusion the premise that there should be sharing, through markets, of standard-of-living risks. In doing this, I proposed quite a number of new markets. But not all of the details of the methods of establishment of these markets have been worked out here. In the absence of a complete specification of the characteristics of the markets and associated institutions, some of the proposed markets appear perhaps unrealistic or improbable. One suspects that the potential of these markets could not be fulfilled, at least for years, even decades.

The development of the new markets proposed in this book, if it is to occur, would plausibly take place in two stages. In the first stage, the users of the markets would be a relatively small number of sophisticated traders who make limited use of the markets for their special purposes. For example, some first users of the residential real estate futures markets might be professional investors in home mortgages and mortgage securities, who use the futures markets to hedge the risk of default and early prepayment that is associated with real estate price changes.[1] First users of national or regional income markets might include managers of firms whose business is particularly sensitive to the income level in their region of business, or managers of investment portfolios who use observable-factor models of returns as part of a risk-management strategy, replacing the national income factors currently in some of these models with their more up-to-date market price.[2] This first stage is important since it promotes the initial liquidity of the markets, creating a base from which use of the markets might grow. In the second stage, these markets would then be used by the general public to protect their standards of living. In this stage, individual homeowners would (with the help of various retailers) use the markets to protect the value of their homes, and individual workers would use the markets to protect their paychecks. The second stage would not come until we have the associated institutions, retail markets, to facilitate the use by the general public of the new markets.

While the first stage use of some of the markets proposed here might not be far off, it is the second stage, when there is broad use of the markets to protect standards of living, that is most important to achieve. To make the full potential for these markets a reality, doubts would have to be overcome, data collected, institutions changed, incentives provided, and the public educated about the merits of the new institutions.

Coming to grips with doubts

There are those today who would advocate the elimination of many of the financial markets we have today, rather than the establishment of major new ones. These people feel that the speculation encouraged by these markets may distract people from more important things, and create wild price fluctuations and financial disorder. These same people might counsel against hedging in some of the new markets.

To someone who has read my past work (1981; 1989; 1992) on the volatility of financial markets, and on the fads and fashions that often interfere with the proper functioning of these markets, it might appear likely that I would be among those who are skeptical of new markets, and I may appear inconsistent in proposing them. There is really no inconsistency, however. I have never advocated the elimination of any existing financial market, so there is no reason why I should oppose the establishment of new such markets.

The decision whether to establish new macro markets should be regarded as essentially the same as the decision whether to allow existing speculative markets to continue. It is hard to see any reason why speculative markets should be allowed, for example, for claims on corporate dividends and not for claims on other components of incomes. No one would seriously propose passing a law that people cannot trade their shares just because the stock market goes through booms and busts. It makes no more sense to say that people should not be able to trade their own income flows for a more stable one.

The observed tendency of speculative markets to boom and crash from time to time for no apparent reason is, however, a cause for concern about all speculative markets, and it is inevitable that we will want to air these concerns before any major new markets are established. Presumably these new markets would sometimes boom and crash too, and there would probably be occasional disruptions caused by these.

In my book *Market Volatility* (1989), after surveying my own research and the research of many others, I concluded that there appears to have been little relation over the past century in the United States between aggregate stock market prices and information about either subsequent future dividends or interest rates. Although the evidence is not without some ambiguities, it appears that aggregate prices of the entire US stock market have been dominated by extraneous volatility.[3] Still, in that same book I reported that I was unable to find evidence of excess volatility in other principal components of US stock prices. Moreover, in that same book I concluded that long-term bond prices do appear to have some (imperfect) ability to predict future interest rates.[4] On the basis of a large amount of evidence in the finance literature, there is now little question but that many speculative asset price movements carry such important information about the relevant future that the movements should not be regarded as dominated by market-generated noise.

It would seem, by analogy, that aggregate-income perpetual claims or perpetual futures markets, if they were established, would have some ability to reveal information about future income flows. It is perhaps likely that, in times when little information about future income is arriving, and at times of investor fads, perpetual claims or perpetual futures price movements may be dominated by destabilizing speculative pressures. On the other hand, in times when there *does* appear to be important new information about the long path of future incomes, and when excessive investors' attention is not attracted by a major fad, the perpetual markets will probably react somewhat appropriately to this information. The same conclusion might be made for other of the proposed markets here. Prices in futures, option or swap markets in real estate would at some times appear to be erratic, while at other times to carry genuine information about future rents (or about the value of future service flows) of real estate.

It has been claimed that, in our existing stock markets, changes in the share price of firms have only a small impact on the firms' investment decisions.[5] This claim, if correct, suggests that speculative markets like the stock market do not discharge much of their supposed responsibility for directing economic resources. However, even if this claim is correct, the reason a firm's own stock price does not influence its own investments more may in part be because the character of firms' new investment decisions tends to bear little relation to the payoff of their own past investments. When we set up perpetual claims or perpetual futures markets, we must be sure, as emphasized in preceding chapters, that the markets price something that can be thought of as a consistent asset or the payoff of an important class of investments, so that the price discovery afforded by these markets would be widely useful.

It is important to remember, moveover, that even if price movements in long-term markets in aggregate incomes, such as perpetual claims or perpetual futures markets, are seriously contaminated by extraneous market-generated noise, the markets can still serve their purpose of allowing people to hedge their income risks. One interpretation of the functioning of such noisy hedging markets would be that the information that is being revealed by short-term movements in the market prices is information about the markets' willingness to take on specific kinds of risk. Information about that willingness is possibly as important as information about the future income flows that are discounted in these markets.[6] Moreover, so long as extraneous noise has only a transient effect on price, that is, so long as price deviations in the perpetual markets do not grow out of bounds, then they may have relatively little impact on a long-term hedger, just as stock market volatility may have relatively little impact on long-horizon stock market investors.

We need also to put this potential problem of irrational speculation-induced price movements in the proper perspective, by comparing it with analogous existing problems. The basic issue is not whether such markets will have irrational speculative bubbles, but rather whether the establishment of these markets will make the sum total of speculative activity any worse. There are already markets where people can speculate today, and there are already speculation-induced price movements. For example, the existing illiquid market for real estate already shows dramatic evidence of speculative instability. Moreover, people today make essentially speculative decisions even in the absence of any ability to trade in markets resembling some of the proposed macro markets. These speculative decisions include decisions whether to invest in new businesses, whether to locate businesses in a given country or region, whether to invest in human capital to go into a certain labor market. A firm today may locate operations in another country because of some fears about the aggregate economic situation in its own country, fears that might better have been dealt with by hedging those risks. Someone today might, let us say, forgo a career in medicine because that person is bothered by risks that occupation might not be lucrative, even though that person has a comparative advantage in medicine. These decisions, to go into a certain occupation, to locate in a certain area, or to make other unmarketable investments, are vulnerable to fashions and fads just as are decisions to invest in liquid assets; both are ultimately made with regard to the unseen future; both are subject to deep ambiguities that may sometimes be resolved in undesirable ways by social-psychological forces. Since the impact of the establishment of macro markets on speculative pressures is ambiguous, it is likely that the hedging abilities that they allow is reason enough to advocate their establishment.

Any doubts about the wisdom of establishing new macro markets should be weighed against the potential benefits. The institution of new markets in aggregate incomes might have many benefits beyond the immediate hedging function. One might say that the establishment of the markets has externalities through their effects on expectations and on other institutions. The markets might fundamentally alter (let us hope improve) the nature of the business cycle. Much macroeconomic theorizing in the tradition of Keynes (1936) has stressed the extreme precariousness of the basis of knowledge on which our decisions must be made, and the instabilities created when markets are confined primarily to current goods. Surely, the absence of markets that would allow expectations about future income flows to be expressed in today's prices is fundamental to macroeconomic dynamics (see, e.g. Arrow and Hahn, 1971).

Some major business cycle theories have been based on rigidities in prices as sources of macroeconomic fluctuations, particularly rigidities in the price of labor. It has been argued (Taylor, 1980) that the tendency of firms to sign contracts with their employees that have a fixed duration (typically one to three years) is a source of persistence to macroeconomic fluctuations, since the contract length imposes rigidities for this interval of time. The duration of other kinds of contracts may also be important for macroeconomic fluctuations (see Fischer, 1977). It is argued that the length of the business cycle (or rather, the frequency distribution of times between recessions) is fundamentally related to the length of these contracts. Long-duration contracts, such as those that specify a wage rate for the duration of the contract, entail a sort of externality: firms that set these contracts do not take into account the effects that their contract has on everyone else, via the effects of the contract on the macroeconomy (Ball, 1987). If there were liquid markets in claims on aggregate incomes, then there might be a fundamental alteration in this tendency towards persistence in macroeconomic fluctuations. Both management and labor would have removed from them some incentive to make rigid contracts. Rather than sign a contract that fixes wages, they could allow wages to move with the market and hedge their risk of wage changes.[7]

Probably the most important prospective benefit of the proposed macro markets is that such markets ought to start a process tending to reduce inequality of incomes. To the extent that Barro (1991) and Barro and Sala-i-Martin (1992) are right, that incomes of nations or regions tend, absent any shocks, to converge to a common mean, then the establishment of markets in national incomes ought eventually to have a major impact on inequalities across nations. Ideally, the markets would even eventually largely eliminate inequalities across nations. Hedging has no effect on predictable consequences. To the extent that the alleged tendency for convergence of nations' incomes to a common mean is predictable, then this tendency for convergence will be unaffected by any hedging of risk. The hedging of risk will instead eliminate the impact of random shocks to income paths, shocks that prevent the convergence of incomes from ever levelling incomes. Macro markets defined over aggregates other than nations may also, of course, ultimately help reduce income inequality across other groupings of people, besides nations, for which macro markets might be defined.

One might hope that people will consider that the inequality of incomes that is routinely endured around the world today, to the extent that it is generated by purely random forces, is an injustice of the highest order. From this perspective, the potential hazards of the new markets seem small.

Providing public goods

If it can be concluded that the future establishment of some of the major macro markets discussed here is worth serious consideration, then it is important now to recognize the extent of the investment in public goods that will be made, and the long time that it takes to create such goods.

Creating a new market requires resources, in such things as creating data sources, in setting up the facilities and institutions for trade in the market, experimenting with market types, publicizing the market, and educating the public about its uses. Resources will not be spent in a free-market economy unless there is a prospect of return on these expenditures. The problem of course, is that, once a market is established, others, who did not share in the costs of the establishment of the market, will reap much of its benefits. This means that there is a tendency for under-investment in the creation of new markets.

In one study of financial innovations (Tufano, 1992), 58 innovative securities were studied that were first issued in the United States between 1974 and 1986. These included various kinds of mortgage-backed securities, asset-backed securities, convertible and non-convertible bonds, preferred stocks, and equities. In all but one of these, a rival entered with innovative products within 78 days of the pioneering issue. The pioneer underwrote an average of only 1.8 of the pioneering deals in the first year since the innovation; there is hardly a long innovation advantage. Moreover, prices charged for their services by the pioneers tended on average to be *lower* than prices charged by rivals; there is no monopoly advantage to pioneers. The only indication of a possible advantage for innovators found in Tufano's study was that there tended to be a larger market share for innovators. The larger market share might be translated into higher profits, even though the price for services charged was lower, if there are economies of scale in underwriting innovative securities; but such economies of scale are hard to measure. Of course, pioneering new products may also benefit the innovator in terms of greater prestige, which may contribute to greater sales in other lines of the business.[8]

Of course, new markets are created with some regularity; that this is so is abundantly apparent in looking at the progress in our financial markets around the world over the last few decades. This fact does not deny the basic principle: new markets tend to be established where the costs and risks are relatively low to establish them, and the costs and risks probably tend to be lowest for minor extensions of existing markets. This means also that innovation may be sluggish, failing to move swiftly into the most important new directions, and occurring only when some accident of history strongly favors the creation of the requisite public goods.

By making public expenditures to support establishment of new markets, we would not need to rely on accidents of history to create such public goods. The establishment of new markets, because of the public nature of its benefits, is one of the most fitting places for government subsidy. Subsidy could take the form of public information and education campaigns, and rewards to successful market innovation. Government regulators can also encourage hedging of risks by organizations.

Developing index numbers

If we decide that the new macro markets are a serious possibility for the future, then new research on index number construction should begin now. A major barrier to the establishment of many macro markets today is a shortage of indices that are agreed upon as well suited to be the basis of cash settlement of contracts. Most statistical agencies have not seen it as their mission to produce indices that are suitable for settlement of contracts.

It is important, before people will sign a contract settled in terms of an index, that there be some history to the index, enough history to enable people to use their judgment, in conjunction with other things that they know about history, to evaluate the indices from many perspectives. The indices must be developed sufficiently in advance of the opening of any new markets to allow time for some debate or discussion about their validity. People must have some time to view the indices as they evolve in real time, to decide how they correlate with their own risks, so that they can hedge effectively.

This history of economic indices shows that such indices are usually begun initially as theoretical constructs, of interest only to a few specialists and economists; national incomes, money stock measures, consumer price indices, and stock price indices all started out as such esoterica. As time passes, after the public grows familiar with these indices and learns what authorities think these indices mean, the indices begin to be regarded as tangible reality and people begin to think in terms of these measures. It is only after this happens that most people might naturally seek out risk-management contracts settled in terms of the indices.

Getting a head start on the development of indices is especially important, since proper construction of some indices may require collection of new data, such as new hedonic variables. Thorough consideration of index number construction methods now may thus result in new data-collection initiatives. The sooner such initiatives are made, the longer the history of indices that will be available. The ability to collect new data will relieve us of the handicap of having to make do with data sets that were collected for other purposes.

It was stressed above that we want our indices to represent prices of or incomes accruing to representative claims on future income streams. Doing that well means collecting panel data, data that follow individual claims through time. Our current national income accounts do not follow such a concept in their data-collection method. Some panel data sets, such as the Panel Study of Income Dynamics in the United States, do try to follow individual incomes through time. Consideration now of how to construct better national income measures may result in an expansion of such panel data-collection efforts, some new series collected, and possibly some changes in methods. Our current producer and consumer price indices do not follow such a concept in their data collection either; they price newly produced goods. Before markets could be opened in durable-goods components of the producer price index, there should be a data-collection effort to follow prices or incomes of specific items. There do not now appear to be good series representing incomes of specific investments in unincorporated businesses; collecting such series could be made possible by the institution of new reporting requirements for a sample of such businesses.

Starting production of new indices means funding researchers to start developing indices with cash settlement of contracts in mind, so that a more secure professional consensus can be reached on how these indices should be constructed. If there is a public recognition of the importance of indices for contract settlement, then the incentive for researchers to reach such a consensus will be improved.

Providing information and incentives

Experimentation with and use of new markets can be fostered by providing public information and creating incentives to trade. Those who provide risk management services have long recognized the importance of educating the prospective customers in proposed new contracts, and know that people may require incentives initially to accept new contracts.

The major futures and options exchanges maintain public relations staffs that provide information, conduct seminars, and produce literature about the use of markets. Starting any new market requires more than just announcing a contract and awaiting buy and sell orders. People unaccustomed to using the contract do not know that they ought to expend the resources to learn about how to use it.

The major exchanges also know that incentives to trade can be especially important for innovative contracts. For example, in 1986 the Coffee, Sugar and Cocoa Exchange, in its attempt to promote its futures contract in the Consumer Price Index, announced that it offered to pay traders $5 a contract for the first 20 contracts traded per day, and that it would create an additional $50,000 pool each month to be distributed to players, on a pro-rata basis, who trade more than 20 contracts a day.[9] Such publicly announced incentives should not be confused with secret efforts to augment trading volume, such as the incentives for cross trading or dummy trading that were allegedly provided by the London Futures and Options Exchange in its 1991 attempt to get property futures markets started.[10] Secret incentives may give potential market participants a false impression of market liquidity. But, so long as the incentives are publicly announced, the incentives benefit everyone. The incentives may be regarded as a legitimate compensation to those people who are providing what would otherwise be a public good, the establishment of a new liquid market.

Launching really innovative new contracts, such as perpetual claims or perpetual futures, may require some intensification and alteration in the procedures for providing information by the institutions that do the launching. Such changes are especially important when the contracts are settled on the basis of income flows, such as national incomes, for which a cash market price has never been observed. Futures exchanges might borrow some methods from underwriters, who are good at finding new markets and establishing new prices. These underwriters undertake publicity programs, establish ongoing relations with clients, keep in communication with these clients as sources of information about possible market prices, and stake their own reputation to some extent on the fairness and reasonableness of the offering price.

Governments may be well advised to provide incentives and information for people to participate in risk-management markets. Governments have a legitimate role in doing this, since the government is ultimately responsible for the well-being of its citizens, so that failure of some people to hedge may create a tax burden for others. Moreover, failure of some people to hedge their risks creates other externalities on others when those failing to hedge are affected adversely by their unhedged risks.

There are already many examples of government incentives to public participation in risk-management contracts. Of course, government management of health insurance is now widespread; it is widely recognized that society cannot bear to allow those who have not insured themselves from suffering all the consequences of ill health. In many cases, such as President Clinton's new national health insurance initiative, government participation has taken the form of requiring institutions or individuals to purchase insurance from private sources. People and organizations are also required to buy flood, automobile liability, and other insurance, so that the government will not have to bear the costs of relieving the suffering that may be caused should accidents occur.

Following such examples, firm's pension funds might be encouraged by regulators or tax authorities to hedge some of the income risks of their beneficiaries. Government insurers of depository institutions might encourage these institutions to hedge their real estate risks. Tax laws might be changed to give individuals, businesses, and farmers incentives to hedge against the price risk on their properties.

Dealing with loci of control

An important obstacle to the establishment of important new risk-management contracts is an institutional rigidity that favors the status quo. Firms may be initially reluctant to offer income insurance policies as employee benefits. Private insurance companies may be skeptical about attempting to market such policies. Mortgage originators may be slow to market home mortgages with home price insurance attached. Banks, savings and loans, and other financial institutions may be slow to hedge their real estate risks. Pension fund managers may be cautious about assuming the responsibility of taking positions in macro markets.

An important reason for sluggishness of response from these institutions is a separation of control over their actions. There are three basic loci of control that operate to guide the activities of these institutions: managers of the institutions, clients of the institutions, and government regulators. The managers are in charge of the day-to-day decisions, but they are under substantial constraints from the other two loci. These three separate loci of control find it difficult to arrive at a consensus as to unusual new steps, such as the adoption of risk-management policies.

In some cases, the clients of institutions are other institutions, corporations, state and local governments, and nonprofit institutions; in other cases the clients are the general public. Either way, the clients tend to be difficult for managers to communicate with. The managers may have the opportunity to present their case to another professional who represents another institution, but decision making at the highest levels in client institutions is likely to be by less professionally expert individuals, and the ultimate control over risk-management decisions may be diffused over many individuals. For example, state legislators may get involved in criticisms of those who manage their pension funds, and the investment managers can hardly hope to be able to present their case to them as they might to a single other investment professional.

Administrators of government regulatory agencies tend to be lawyers or investment professionals with substantial expertise about financial markets, but, again, the costs and time it takes to convince them of the wisdom of some financial innovation is another obstacle on the way to innovation. Moreover, the administrators are themselves subject to oversight by others, and the political process affects their decision making as well.

Problems coordinating these loci of control have delayed many financial innovations. Ten years elapsed between the trading of the first financial futures (the currency futures first traded at the Chicago Mercantile Exchange in 1972) and the first stock index futures (the Value Line contract at the Kansas City Board of Trade in 1982). Years elapsed between the establishment of the Treasury Bond futures market (on October 20, 1975) at the Chicago Board of Trade before there was a substantial amount of trade. The Consumers Price Index futures contract, first proposed by Lovell and Vogel in 1973, and thereafter soon receiving widespread endorsement by economists, was not traded until 1985 (at the Coffee, Sugar, and Cocoa Exchange in New York).

Dealing with separate loci of control means essentially getting their attention together using major public media. There must be a very public discussion of the issues, not just an effort to deal with these loci of control individually.

The importance of opinion leaders

The general public does not have to be convinced of the importance to them personally of fluctuations in national income and other income aggregates; they already are. Indeed, national elections for public office typically center on what the candidates will do for such income aggregates. Daily newspapers regularly focus attention on indicators of future income, even though no speculative market exists in such incomes.

But this concern with aggregate income fluctuations does not translate immediately into any public interest in hedging markets or retail products that allow people to insure themselves against the risk of fluctuations in income. The reason it does not is that the public has not made the connection that it is feasible and wise to insure against fluctuations in such risks.

The public at large will not buy an insurance product based only on an intellectual argument, made to them directly, that such products are in their interest. Opinion leaders (writers and commentators, investment and tax advisors, financial planners, lawyers, regulators, and lawmakers) would have to come to some professional consensus on the importance of hedging major economic risks, and on how properly to design and use hedging contracts, before the use of most of these macro markets naturally became widespread. These people would have some work to do before attaining such a consensus. They must study and understand the features of the contracts and associated laws and institutions. Such opinion leaders would have to come to some agreement that the costs of risk-management contracts (in the form of expected backwardation or option premia) are sufficiently low to mean that individuals are indeed better off hedging, and to communicate this consensus to the public. With risks that may be perceived to have a low probability of turning out to be catastrophic, people will be sensitive to the perceived costs of insurance, and these costs may be difficult for them to judge by themselves.

Individuals do not need to understand or assimilate all of the reasons for the economic institutions that they have in order to use these institutions appropriately. What must be developed to allow economic development to proceed is more social than individual: the further development of authorities, and public recognition of these authorities, including authorities on the technology of markets who maintain a body of thought on the proper functioning of markets and proper hedging procedure.

The psychological impulses that may inhibit individual initiative to hedge risks can coexist with enlightened hedging behavior, and, also, perhaps, with a personal sense of satisfaction and relief that risks are managed appropriately. Achieving a state of public opinion that would enable such risk management could be regarded as another step along the way to greater economic development, like the many advances in public opinion before it that have made our current quality of life possible.

Notes

Chapter 1

1. The closest I have been able to find in any published source to the national income markets proposed in this book is an article advocating swaps and options on business cycle variables, particularly indices of consumer confidence, Marshall *et al.* (1992). The Coffee, Sugar and Cocoa Exchange in New York (1983), under the tutelage of Todd Petzel, established an Economic Index Market Division to trade an array of index numbers, including the consumer price index, corporate earnings, housing starts, and new car sales. Only the consumer price index futures was actually attempted.
2. There is the problem that purchasing power may not be transferable on a one-to-one basis. Avinash Dixit has pointed out to me that the macro markets may not function well, for example, to protect people in a country against adverse shocks to their own non-tradable sector.
3. Less developed countries have some different, informal, means of sharing income risks at the village level, see Rosenzweig (1988) and Townsend (1993). But the sharing is only local, and does not allow effective sharing of risks about regional or national aggregates.
4. The Securities and Exchange Commission, set up to encourage liquid stock markets, did so with a number of policies: such things as 'uptick rules to deter concerted activity by short sellers; 13(d) rules to prevent undisclosed groups from acting in concert; extensive reporting and auditing requirements for corporations; Section 144 and 16(b)5 regulations on stock sales by executives and insiders' (Bhide, 1991).
5. The cause of the CPI futures market was taken up by Petzel (1985) and Petzel and Fabozzi (1986).
6. The US CPI experiment was also compromised (as will be argued below) by a contract design that was too short-term in maturity, and once the contract failed, there was little willingness to experiment again.

7. The government obligations on whose payouts the futures contracts were settled were called Obrigações do Tesouro Nacional (OTN) and, later, Bônus do Tesouro Nacional (BTN). These obligations were indexed to the Índice de Preços ao Consumidor (IPC). See Garcia (1991).
8. Four contracts in all were launched at the same time. There was also a mortgage interest rate contract and a rent contract (the latter a conventional, not perpetual, futures contract).
9. See 'Fox on the rocks' and 'Fox rocked by resignations', *Futures and Options World* (Nov. 1991), 9–11.

Chapter 2

1. There is also another way that failure to purchase insurance might be reconciled with prospect theory. In that theory, people do not use their actual probabilities in the decision process; they may round low probabilities to zero; see Kahneman and Tversky (1979). Moreover, appending to the theory an endowment effect, Knetsch *et al.* (1989), a tendency to prefer what one has over alternatives, may help explain why people might tend to accept the risks they have.
2. There are some curious patterns of human behavior in the face of ambiguity, that do not seem rational. Consider the behavior pattern widely known as the 'Ellsberg paradox'. Ellsberg found that, if subjects were told that they would be given a reward if a ball of one of two colors were drawn from an urn, and they were allowed to choose between an urn with known proportions, 50% of both colors, and an urn with an unknown proportion of the two colors, they tended to choose the urn with the known proportions. This behavior pattern, called ambiguity avoidance, appears irrational, since the lack of knowledge about both urns is really the same. How much we should generalize such behavior patterns to situations of careful, social, decision making is unclear.
3. See Grossman (1977) and Arrow (1974).
4. If people's demand is determined by a long weighted moving average of past price changes, then the feedback loop of an irrational speculative bubble may occur without much serial correlation in price changes; see Shiller (1990). The presence of some irrational 'noise' traders may cause rational traders to destabilize markets, see Cutler *et al.* (1990) and De Long *et al.* (1990*a*; 1990*b*).

Chapter 3

1. Concern about income in the distant future does not imply that hedgers will want long maturity futures contracts: so long as the futures contract is cash settled based on an asset price, then the horizon of the futures contract is effectively perpetual in terms of income even if it cash settles next month.
2. For example, a person desiring to hedge only five years' income could combine a long position in an income perpetuities market with a forward contract for the sale of an income perpetuity after five years. Such forward contracts could be arranged on an individual basis once markets for perpetual claims are established.
3. See Kupiec (1990) for a description of these instruments.
4. Both shorts and longs would be obligated by the exchange also to maintain margin accounts, and both shorts and longs could earn interest on these accounts. Equation (3.1) shows the total flow from shorts to longs.
5. The law of one price has, as will be discussed below, its exceptions, e.g. closed-end mutual funds whose price does not correspond to the value of the constituent shocks.
6. Tirole (1982) has shown that bubbles in prices of conventional securities rely on the myopia of traders and disappear if traders adopt a truly dynamic maximizing behavior. The rational bubbles may be contrasted with the irrational bubbles discussed in the last chapter, which would eventually burst.
7. 'People who argue that speculation is generally destabilizing seldom realize that this is largely equivalent to saying that speculators lose money, since speculation can be destabilizing in general only if speculators on the average sell when the currency is low in price, and buy when it is high' (Friedman, 1953: 175). He is writing here about stationary, not explosive, speculation-induced price movements.
8. Blanchard and Watson (1982) showed that there may be bursting bubbles that are nonetheless rational, however in their example the expected price still diverges.
9. See, for example, Black (1986), and Lee *et al.* (1990). The closed-end mutual fund example analogy may not be apt, since there is more potential reason for price discrepancies with the mutual funds: people may have varying trust in the quality of the management of the mutual funds.

10. This equation can also be derived more directly by a no-arbitrage argument.

Chapter 4

1. According to a theory in Friend and Blume (1975), the market risk premium, defined as the expected return on the market portfolio minus the risk-free rate, should equal the coefficient of relative risk aversion times the variance of the market portfolio. They did not have the variance of the true world market portfolio, a number that will be estimated later in this chapter.
2. Since it will not be possible for individuals to sell claims on their children's income, there is an inherent limitation on their ability to protect their minor children against adverse income shocks.
3. Of course, there would likely also be markets for puts on macro market prices, which would facilitate such partial hedging, and eliminate the risk that sudden jumps in price would vitiate the dynamic hedging strategy.
4. This fact was first documented by Feldstein and Horioka (1980). Subsequent research has provided further confirmation, see for example Obstfeld (1993*b*).
5. The Coffee, Sugar and Cocoa Exchange (1983) briefly considered starting a futures market in a corporate-earnings index, though not a perpetual futures market.
6. The employment cost indices, for example, those produced by the US Bureau of Labor Statistics, might be improved if they are to be ideal for settling contracts. They should ideally be put on a repeated-measures basis, following individuals through time. Moreover, labor costs have many intangibles to them; the direct wage costs are only a component of them. There are also the costs involved in worker compliance and cooperativeness. For the purpose of helping firms to hedge their employment costs better, some effort might be made to include such costs in a labor cost index. A good way to measure such labor costs might be to run a regression, with some measure of distress or success of firms as the dependent variable, and various indicators of cost as right-hand variables. The measures of distress of firms might be measured less frequently than the right-hand variables; only the right hand side variables need be measured every time an index value is announced.

7. 'Worried doctors across the country are selling their offices to investor-owned public companies, hoping to bolster their prospects in the rapidly changing health care system. The doctors are signing long-term contracts to work with the companies in some cases as long as 33 years.... Companies like Caremark typically give the doctors cash and stock and promise them a steady income...', Milt Freudenheim, *New York Times* (Sept. 1, 1993), 1.
8. Campbell (1993) has produced a similar estimate of the variance in the United States of returns in a market for a perpetual claim on future labor income.
9. The conspicuous omissions from this table are the USSR and China, for which the Summers and Heston data did not extend back to 1950.
10. The same value of ρ was chosen here for each country, even though different countries have historically had different growth rates, which would suggest higher ρs for the higher growth rate countries. Implicitly I am assuming here that the risk premium in the discount rate is higher for higher growth rate countries. Of course, estimated volatility could be made larger or smaller by lowering or raising $\hat{\rho}$.
11. The real GNP series used here is produced by linking to the US gross national product in 1987 dollars divided by the US population to the Kendrick real per capita GNP in 1958 dollars (US Dept. of Commerce, Bureau of Economic Analysis, 1973). There has been some controversy whether the apparent greater volatility of the pre-war US gross national product may be an artifact of earlier data collection procedures (Romer, 1989; Balke and Gordon, 1989).
12. The values shown in Table 4.1 are converted for all countries from the 1985 base year used in the Penn World Table to the 1990 base year using the US implicit deflator for gross domestic product (1987 = 100).
13. The present values for some foreign countries may seem especially high to those accustomed to using exchange rates to convert foreign incomes to dollars. One of the most important—and stressed—results of the UN International Comparisons Project that gave rise to these GDP figures is that such exchange rate conversions do not give anything like an accurate sense of relative real GDPs (Summers and Heston, 1988; 1991).
14. A complicating factor in trying to judge the extent to which prices in macro markets might correlate with prices in existing financial markets is that there might be some correlation in prices even if

there is no correlation at all between the aggregate income in the macro market and aggregate dividend series in the existing financial markets. One way in which this might come about is information pooling (Beltratti and Shiller, 1993). An information variable may exist that reveals, say, the sum of the aggregate income series for the macro market and the dividend series for the financial market. Negative information pooling could conceivably eliminate any correlation in price changes between the two markets even if the aggregate income and dividend series are positively correlated.

15. Brainard and Dolbear (1971) pointed out the low correlation of corporate profits with national income, and noted the significance of this low correlation for the allocation of social risk. With more recent data, the correlation is higher: the correlation between the two series plotted in Fig. 4.2 between 1964 and 1992 is 41.58%. The correlation between the five-year growth rate in real per capita dividends and the corresponding growth rate in real per capita national income from the US National Income and Product Accounts 1964–92 is 48.30%.
16. Obstfeld (1993*a*) finds that some of the lack of correlation of consumption across countries is due to differing responses across countries to oil price shocks; if this is right, some of the consumption risk could be hedged in oil futures markets.
17. To see this, note that ξ_t is $-\rho$ times the innovation at time $t + 1$ in δ_{t+1} (this innovation, from (4.7), equals $-e1'A(I - \rho A)^{-1}u_{t+1}$) plus the innovation at time $t + 1$ in Δd_t (this innovation equals $e1'u_{t+1}$).
18. The dividend series used here starting in 1926 is dividends per share, 12 months moving total adjusted to index for the last quarter of the year, composite, as reported by Standard and Poor's Statistical Service. Values before 1926 were created by linking the Standard and Poor's series to a series produced by Cowles (1939). These nominal dividends were converted to real dividends by dividing by the producer price index 1982 = 100 for Jan. of the succeeding year. The linked dividend series appears as Series 2 and the producer price index as Series 5 in Shiller (1989: ch. 26).
19. Campbell and Shiller (1988; 1989), using similar methods, estimated that the stock returns were about twice as volatile as the present-value model with constant discount rate would imply.
20. Since the variance matrix Ω for the error in the autoregressive model for real dividends is estimated with the usual degrees-of-freedom correction, the improvement of fit of the regression caused by the

addition of explanatory variables does not directly reduce the estimated variance of the error, as the number of explanatory variables is increased. Of course, the possibility of more subtle small sample biases in this estimate remains.

21. This suggests that the market overreacts to dividends relative to the present-value model, as has been noted before (see Shiller, 1989).

Chapter 5

1. These estimates, based on property taxes paid and the inferred assessed values, and on corrections of assessed values to market values, are from Miles *et al.* (1991: Exhibit V).
2. According to the *Balance Sheets for the US Economy* (US Board of Governors of the Federal Reserve System), in 1990 the value of residential structures in the USA was $4,796 billion, and the value of land was $4,940 billion; these alone made up more than half their estimated domestic wealth of $18,228 billion. Other estimates of real estate wealth are discussed in Miles *et al.* (1991).
3. These data are somewhat unreliable: a simple median is vulnerable to changes in the mix of housing sold, but the data are good enough to give us an indication of the variability of prices over moderately long intervals.
4. These figures on Japan and Korea are from Kim (1993).
5. Some of the short-run wiggles in the real estate price index may be regarded as due to sampling error. The index is based on the universe of sales, but we may consider the sales observed as a sample of the value of all houses. Standard errors for the level of this index are generally less than 1%. Sampling error imposes an approximately serially uncorrelated noise on the index, making the curve plotted in the figure look like a smooth curve drawn with a trembling hand. (See the discussion of the equal representation model in Ch. 6 below.)
6. The stock price index shown in Fig. 5.2 does show an uptrend over this sample period; the sample corresponds to the bull market of the 1980s and early 1990s, interrupted at various points, notably the stock market crash of 1987. Random walks will often show such overall uptrends in finite samples just by chance, but they are not likely to show smooth uptrends. Because of the interruptions the uptrend in Fig. 5.2 is not smooth.

7. Noguchi (1992*a*; 1992*b*) has proposed establishing 'land value index bonds' to securitize Japanese land; suggesting that such bonds might mitigate the tendency for land price speculation in evidence in Japan.
8. Existing commodity swaps and commodity-linked debt share some of the benefits of such contracts, though only for a finite horizon.
9. McCulloch (1980) argues that index bonds were illegal in the United States from 1933 to 1977; there are no legal barriers now. See Hochman and Palmon (1988), and Knoll (1991). For the UK barriers, see Bootle (1991).
10. The term 'money illusion' was created by Fisher (1928). That money illusion infects human decision making has been shown experimentally, see Diamond *et al.* (1992).
11. According to the US Bureau of Labor Statistics, the standard errors of month-to-month changes in the national Consumer Price Index are quite small relative to the standard deviation of the month-to-month changes. However, standard errors of short-run changes in city indices or in components of the national index are high enough to raise serious questions about the meaning of the measured short-run changes.
12. Bodie (1990) has detailed the potential advantages of long-horizon inflation-protected annuities and long-maturity inflation options.
13. Agricultural perpetual futures markets that use an interest rate indexed to the general price level as the alternative asset return for settlement would also have the advantage, relative to existing agricultural futures markets, of locking in for the farmer the real, rather than nominal, price.
14. Farm prices, just as other speculative asset prices, appear to show more volatility than one might expect based on fundamentals; see Falk (1991).
15. There have been limited partnerships in art and collectibles, which maintain collections for investors, and these do increase liquidity for investors. Still, these are not liquid enough to function as useful hedging media. They also appear vulnerable to the same departures of price from asset value that we see in closed-end mutual funds. For example, in 1993, Merrill Lynch and Numismatic Fine Arts International liquidated their Athena II limited partnerships in rare gold and silver coins and ancient art, originally valued at $25 million, after units could not be sold for a fraction of the original price. Such limited partnerships might be expected to do poorly for

another reason: the investors do not receive the dividend of possessing and enjoying the art.

Chapter 6

1. In constructing stock price indices, one must confront the problem that prices of new issues of shares need not be the same as market prices; with 'rights' issues they may differ markedly. The 'base-weighted aggregative' formula used to compute the Standard and Poor's stock price indices reduces to the value-weighted arithmetic index described here, equation (6.1), if these prices are the same; see Cowles (1939: expression (18.2)). Standard and Poor's also restricts its sample to 500 stocks, although the effect of this restriction is small since these stocks constitute most of the value of the market.
2. The best thing to do, in the real estate application, if we have the information on age of properties, may be to use only repeat-sales data to construct indices, thereby excluding new houses that have sold only once, but also use a hedonic repeated-measures index construction method to correct for the mix of ages of properties. These hedonic repeated-measures indices are described in this chapter below.
3. Since people do not normally reinvest their service income in more housing, the index with income reinvested would tend to grow through time relative to a standard claim on income; hedgers would have to adjust downward through time the number of contracts held.
4. It is assumed for simplicity that the list of quality variables is unchanging through time, though there is nothing in these methods that requires this. In fact, as time goes on we are likely to discover new quality variables, and want to discard old ones.
5. We may want to drop properties entirely where there has been a change in floor space, on the assumption that the change in floor space may indicate that there may also have been other changes in quality at the time of the remodelling. Of course, if we drop all remodelled properties, then our index will tend to have a lower rate of increase than if we had not, since part of remodelling is periodic maintenance. Ideally, the term p_{it} should represent, not the selling price of the *i*th property at time *t*, but the cumulated value of the investment, including all improvements made and rents collected for the property, in which case this bias would be eliminated.

6. A suggestion that heterogeneity plays a role in causing the constant term was found by lumping data sets for the four cities from across the country, from Case and Shiller (1987), into one large data set, and estimating a national index. The constant term was 63% larger than observation-weighted average constant terms for the four cities, and bigger than the constant term for any of the four.
7. To understand the potential biases due to heterogeneity, consider an extreme case of departure from assumptions of our model. Suppose that on the west side of a county the real estate market is 'hot', with prices rising at 10% a year and houses are selling every year. Suppose also that on the east side of the county the market is 'cold' with prices rising at 10% only every two years, and that houses are selling only every two years. A Goetzmann–Spiegel regression with these data for the entire county would produce a constant term of 10% and all slope coefficients zero; as a result their index would show no price appreciation at all for the county. In contrast, an ordinary repeated-measures index like that described here would show an appreciation somewhere between 5% and 10% a year, between the actual appreciations in the two sides of the county. In this case, of course, the ordinary repeated-measures index would tend to show an overall increase closer to 10% than 5%, due to overweighting of the hot market in the regression; methods of dealing with this problem are discussed in this and the next chapter. In terms of robustness to heterogeneity with ordinary repeated-measures indices, this example suggests that the regression without the constant term appears to have an advantage over the regression with the constant term.
8. In this example there are no more than two observations of any one property. If there are more than two sales of some property, then we should ideally take a generalized least-squares estimate that takes into account the negative correlation between the error terms corresponding to consecutive pairs of sales of the same property, due to the presence of the middle price with opposite signs in the two error terms. This will be done automatically in the methods presented here if the Ω matrix represents that there is temporary noise in price at time of sale, i.e. if the variance of such noise is added to the diagonal elements of this matrix. Then the variance matrix ω, formed as $S_1'\Omega S_1$, takes such tendency for negative covariances into account. It is not particularly difficult to invert the variance matrix of errors, since it is block-diagonal, a block for each subject (property).

9. If T is greater than 3, then some of these ways of inferring price changes may be correlated with each other.
10. Goetzmann has proposed a method to correct for this bias in a geometric repeat-sales price index using estimated variances and a log normality assumption.
11. If s_{it} changes through time, we may include an additional column in x whose ith element is $-s_{i0}P_{i0}$ if the property was sold at time 0, otherwise 0. (The corresponding column would also be added to z.)

Chapter 7

1. According to the 1991 *Statistical Abstract of the United States*, in 1987 3.53 million existing one-family homes sold (Table 1275) and there were 58.16 million owner-occupied year-round units (Table 1282).
2. The value of s required still appears low. For example, the Dallas metropolitan area contains about 1.5% of the US population, suggesting that there should be about 50,000 sales of existing single family homes per year in that metropolitan area; our 17.5-year-long sample contained 211,000 sales in total, about a quarter of the number of sales we might expect.
3. See Goetzmann (1990) and Kuo (1993).
4. Data on new construction improvements are from US Dept. of Commerce, *Construction Review*, Table B5, p. 12, Nov./Dec. 1991.
5. Doing this may be preferable to including a constant term in the repeat-sales regression, as advocated by Goetzmann and Spiegel (1992), since it takes account of the tendency for improvements to depreciate.
6. Shlaes (1984: 496).
7. Quoted in Marchitelli and Korpacz (1992: 314).
8. The problem that data are missing does not rule out estimating factor models with more than two factors, though we could not expect to deal with more than one quality factor of each house sold only twice.
9. Webb *et al.* (1993) proposed such modelling for the purpose of generating price indices. S. Kim (1992) used a similar method to produce indices of rents in the housing market. Analogous methods may be used to correct for a different potential problem in the construction of price indices: that some of the properties may have

been changed or improved between sales. Montgomery (1990) showed how a qualitative-choice model involving the decision to improve may be used to help estimate a housing expenditure equation.

10. Webb *et al.* (1993) proposed such a two-step procedure. Their method however differed from that discussed here in that their hedonic equation did not allow any time variation in coefficients; time variation in the price index came entirely from time variation in the hedonic variables (which included rent). Two-step procedures are treated in Maddala (1983: ch. 8).

Chapter 8

1. Here, the assumption is that we have the entire history of sales prices, and that the future shows no different sales frequency than the past.
2. Of course, these, and any hedonic methods that omit subject dummies, are not robust to changes in the mix through time of subjects observed, and may not be ideal for producing price indices.

Chapter 9

1. There are substantial effects of residential real estate prices on mortgage default, see Case *et al.* (1993*b*).
2. Chen *et al.* (1986) find that changes in a US industrial production index (a monthly indicator of US national income, which is published quarterly) is a very important factor in stock market returns. They write (p. 402): 'Perhaps the most striking result is that even though a stock market index, such as the value-weighted New York Stock Exchange Index, explains a significant portion of the time-series variability of returns, it has an insignificant influence on pricing (i.e. on expected returns) when compared against the economic variables.'
3. Evidence for excess volatility in the stock markets of the UK and Germany has been less dramatic; see Bulkley and Tonks (1989) and De Long and Becht (1992; 1993).
4. Some of these conclusions were the result of joint work with John Campbell (1987; 1988; 1989).

5. See Morck *et al.* (1990), Rhee and Rhee (1991), and Blanchard *et al.* (1993).
6. This potential for variation in the markets' willingness to bear risk provides another important reason to have long-horizon markets, such as perpetual futures, rather than one-period-ahead income futures. With the long-horizon markets, people can insure themselves against the possibility that the market will not want to bear risk at favorable rates in future years.
7. There may be other macroeconomic effects. Abel (1988) has argued that, in theory, income insurance ought to reduce precautionary saving.
8. Futures exchanges represent a highly competitive industry, see Carlton (1984). An exchange that innovates a new concept can see its advantage stolen away by a competing exchange. Many initially successful new contracts fail later when another exchange finds a slightly better version of the contract.
9. 'Trade trends: CPI paybacks', *Futures* (Feb. 1986), 34.
10. See 'Fox on the rocks' and 'Fox rocked by resignations', *Futures and Options World* (Nov. 1991), 9–11.

References

Abel, Andrew B. (1988). 'The Implications of Insurance for the Efficacy of Fiscal Policy', *Journal of Risk and Insurance*, **55**, 339–78.

Abraham, Jesse M. (1990). 'Statistical Biases in Transaction-Based Indices', unpublished working paper, Federal Home Loan Mortgage Corporation.

—— and William S. Schauman (1991). 'New Evidence on Home Prices from Freddie-Mac Repeat Sales', *AREUEA Journal*, **19**, 333–52.

Akerlof, George A. (1970). 'The Market for "Lemons": Quality Uncertainty and the Market Mechanism', *Quarterly Journal of Economics*, **84**, 488–500.

Allen, B. Paul, Heifner, Richard G., and Helmuth, John W. (1977). *Farmers' Use of Forward Contracts and Futures Markets*, US Department of Agriculture, Economic Research Service, Argicultural Economic Report No. 320, Washington, DC.

Arrow, Kenneth J. (1964). 'The Role of Securities in the Optimal Allocation of Risk Bearing', *Review of Economic Studies*, **31**, 91–6.

—— (1974). *Essays in the Theory of Risk-Bearing*, North Holland, Amsterdam.

—— (1991). 'Risk Perception in Psychology and Economics', *Economic Inquiry*, **20**, 1–9.

—— and Hahn, Frank F. (1971). *General Competitive Analysis*, Holden-Day, San Francisco.

Atkeson, Andrew, and Bayoumi, Tamim (1991). 'Do Private Capital Markets Insure against Risks in a Common Currency Area? Evidence from the United States', unpublished working paper, University of Chicago.

Backus, David, Kehoe, Patrick, and Kydland, Finn (1992). 'International Real Business Cycles', *Journal of Political Economy*, **100**, 745-75.

Bailey, Martin J., Muth, Richard F., and Nourse, Hugh O. (1963). 'A Regression Method for Real Estate Price Index Construction', *Journal of the American Statistical Association*, **58**, 933–42.

Balke, Nathan S., and Gordon, Robert J. (1989). 'The Estimation of Prewar Gross National Product: Methodology and New Evidence', *Journal of Political Economy*, **97**, 38–92.

Ball, Laurence (1987). 'Externalities from Contract Length,' *American Economic Review*, **77**, 615–29.

Barro, Robert J. (1991). 'Economic Growth in a Cross Section of Countries', *Quarterly Journal of Economics*, **106**, 407–43.

—— and Sala-i-Martin, Xavier (1992). 'Convergence', *Journal of Political Economy*, **100**, 223–51.

Beltratti, Andrea, and Shiller, Robert J. (1993): 'Actual and Warranted Relations between Asset Prices', *Oxford Economic Papers*.

Berck, Peter (1981) 'Portfolio Theory and the Demand for Futures: The Case of California Cotton', *American Journal of Agricultural Economics*, **63**, 466–74.

Bhide, Amar (1991). 'Active Markets: Deficient Governance', unpublished working paper, Harvard Business School, Cambridge, Mass.

Bikhchandani, S. D., Hirshleifer, David, and Welch, Ivo (1990). 'A Theory of Fashion, Custom, and Cultural Change', unpublished working paper, University of California, Los Angeles.

Black, Fischer (1986). 'Noise', *Journal of Finance*, **41**, 529–43.

Blanchard, Olivier J., Rhee, Changyong, and Summers, Lawrence (1993). 'The Stock Market, Profit and Investment', *Quarterly Journal of Economics*, **108**, 115–36.

—— and Watson, Mark W. (1982). 'Bubbles, Rational Expectations, and Financial Markets', in Paul Wachtel, ed., *Crises in the Economic and Financial Structure: Bubbles, Bursts and Shocks*, Lexington Books, Lexington, Mass.

Board of Governors of the Federal Reserve System (1992). *Balance Sheets for the U.S. Economy 1960–91*, Washington, DC.

Bodie, Zvi (1990). 'Inflation Insurance', *Journal of Risk and Insurance*, **57**, 634–45.

Bootle, Roger (1991). *Index Linked Gilts: A Practical Investment Guide*, 2nd edn., Woodhead-Faulkner, New York.

Brainard, William (1973). 'Private and Social Risk and Return to Education', in R. Layard and R. Attiyah (eds.), *Efficiency in Universities: The La Paz Papers*, Elsevier.

—— and Dolbear, F. T. (1971). 'Social Risk and Financial Markets', *American Economic Review*, **61**, 360–70.

Breeden, Douglas, (1979). 'An Intertemporal Asset Pricing Model with Stochastic Consumption and Investment Opportunities', *Journal of Financial Economics*, **7**, 265–96.

Brennan, Michael J. (1990). 'Latent Assets', *Journal of Finance*, **45**, 709-31.

Bulkley, George, and Tonks, Ian (1989). 'Are U.K. Stock Prices Excessively Volatile?' *Economic Journal*, **99**, 1083–98.

Cagan, Philip (1971). 'Measuring Quality Changes and the Purchasing Power of Money: An Exploratory Study of Automobiles', in Zvi Griliches (ed.), *Price Indexes and Quality Change*, Harvard University Press, Cambridge, Mass.

Camerer, Colin, and Kunreuther, Howard (1989). 'Experimental Markets for Insurance', *Journal of Risk and Uncertainty*, **2**, 265–300.
Campbell, John Y., (1993). 'Understanding Risk and Return', unpublished working paper, Princeton University.
—— and Shiller, Robert J. (1987). 'Cointegration and Tests of Present Value Models', *Journal of Political Economy*, **95**, 1062–88.
—— —— (1989). 'The Dividend–Price Ratio and Expectations of Future Dividends and Discount Factors', *Review of Financial Studies*, **1**, 195-228.
—— —— (1988). 'Stock Prices, Earnings, and Expected Dividends', *Journal of Finance*, **43**, 661–76.
—— —— (1991). 'Yield Spreads and Interest Rate Movements: A Bird's Eye View', *Review of Economic Studies*, **58**, 495–514.
Carlton, Dennis (1984). 'Futures Markets: Their Purpose, Their History, Their Growth, Their Successes and Failures', *Journal of Futures Markets*, **4**, 237–71.
Case, Bradford, Pollakowsi, Henry O., and Wachter, Susan M. (1991). 'On Choosing among House Price Index Methodologies', *AREUEA Journal*, **19**, 286–307.
—— and Quigley, John (1991). 'The Dynamics of Real Estate Prices', *Review of Economics and Statistics*, **73**, 50–8.
Case, Karl E. (1986). 'The Market for Single Family Homes in Boston', *New England Economic Review*, May/June, 38–48.
—— and Shiller, Robert J. (1987). 'Prices of Single Family Homes Since 1970: New Indexes for Four Cities', *New England Economic Review*, Sept./Oct., 45–56.
—— —— (1988). 'The Behavior of Home Buyers in Boom and Post-boom Markets', *New England Economic Review*, Nov./Dec., 29–46.
—— —— (1989). 'The Efficiency of the Market for Single Family Homes', *American Economic Review*, **79**, 125–37.
—— —— (1990). 'Forecasting Prices and Excess Returns in the Housing Market', *AREUEA Journal*, **18**, 253–73.
—— —— and Weiss, Allan N. (1993*a*). 'Default Risk and Real Estate Prices: The Use of Index-Based Futures and Options in Real Estate', unpublished working paper, Cowles Foundation, Yale University, New Haven, Conn.
—— —— —— (1993*b*). 'Index-Based Futures and Options Markets in Real Estate', *Journal of Portfolio Management*, **19**, 83–92.
Chen, Nai-Fu, Roll, Richard, and Ross, Stephen A. (1986). 'Economic Forces and the Stock Market', *Journal of Business*, **59**, 383–403.
Chinloy, Peter T. (1977). 'Hedonic Price and Depreciation Indices for Residential Housing: A Longitudinal Approach', *Journal of Urban Economics*, **4**, 469–82.

Christofferson, Anders (1970), *The One Component Method with Incomplete Data*, Selected Publications Vol. 25, Department of Statistics, University of Uppsala.

Clapp, John M., and Giaccotto, Carmelo (1992*a*). 'Appraisal-Based Real Estate Returns under Alternative Market Regimes', *AREUEA Journal*, **20**, 1–24.

—— —— (1992*b*). 'Estimating Price Trends for Residential Property: A Comparison of Repeat Sales and Assessed Value Methods' *Journal of the American Statistical Association*, **87**, 300–6.

—— —— and Tirtiroglu, Dogan (1991). 'Housing Price Indices Based on All Transactions Compared to Repeat Subsamples', *AREUEA Journal*, **19**, 270–85.

Clarke, Grindlay (1991). 'A future for property?', *Futures and Options World*, Nov., 20.

Coffee, Sugar and Cocoa Exchange (1983). *Justification for the Coffee, Sugar and Cocoa Exchange's Earnings Index Futures Contract*, NY.

Copeland, Thomas E., and Galai, Dan (1983). 'Information Effects on the Bid–Asked Spread', *Journal of Finance*, **38**, 1457–69.

Cornell, Bradford, and French, Kenneth R. (1983). 'Taxes and the Pricing of Stock Index Futures', *Journal of Finance*, **38**, No. 3, 675-94.

Court, Andrew T. (1939). 'Hedonic Price Indices with Automobile Examples', in *The Dynamics of Automobile Demand*, General Motors.

Cowles, Alfred (1939). *Common Stock Indexes, 1871–1937*, Cowles Commission for Research in Economics, Monograph No. 3, 2nd edn., Principia Press, Bloomington, Ind.

Cox, John, Intersoll, Jonathan, and Ross, Stephen (1981). 'The Relationship of Forward Prices and Futures Prices', *Journal of Futures Markets*, **9**, 321–46.

Cox, Larry C., Gustafson, Sandra G., and Stam, Antonie (1991). 'Disability and Life Insurance in the Individual Insurance Portfolio', *Journal of Risk and Insurance*, **58**, 128–46.

Cutler, David M., Poterba, James M., and Summers, Lawrence H. (1990). 'Speculative Dynamics and the Role of Feedback Traders', *American Economic Review*, **80**, 63–8.

De Alessi, Louis (1987). 'Why Corporations Insure', *Economic Inquiry*, **25**, 429–38.

DeBondt, Werner F. M., and Thaler, Richard H. (1985). 'Does the Stock Market Overreact?', *Journal of Finance*, **40**, 793–805.

Debreu, Gerard (1959). *Theory of Value*, New York.

De Long, J. Bradford, and Becht, Marco (1992). '"Excess Volatility" and the German Stock Market 1876–1990', unpublished working paper, Harvard University.

—— —— (1993). '"Excess Volatility" on the London Stock Market 1870–1990', unpublished working paper, Harvard University.

—— Shleifer, Andrei, Summers, Lawrence H., Waldman, Robert J. (1990*a*). 'Noise Trader Risk in Financial Markets', *Journal of Political Economy*, **98**, 703–38.

—— —— —— —— (1990*b*). 'Positive Feedback Investment Strategies and Destabilizing Rational Speculation', *Journal of Finance*, **45**, 379–95.

Diamond, Peter, Prelec, Drazen, Shafir, Eldar, and Tversky, Amos (1992). 'Money Illusion', unpublished working paper, MIT, Cambridge, Mass.

Divisia, François (1925–6). 'L'indice monétaire et la théorie de la monnaie', *Revue D'Economie Politique*, **39**, 842–64, 980–1001.

Early, John F., and Sinclair, James H. (1983). 'Quality Adjustment in the Producer Price Indices', in Murray F. Foss (ed.), *The U.S. National Income and Product Accounts: Selected Topics*, NBER Studies in Income and Wealth, vol. 47, Cambridge, Mass.

Einhorn, Hillel J., and Hogarth, Robin M. (1986). 'Decision Making under Ambiguity', *Journal of Business*, **59** (No. 4, pt. 2), S225–50.

Eisner, Robert, and Strotz, Robert H. (1961). 'Flight insurance and the theory of choice', *Journal of Political Economy*, **69**, 355–68.

Ellsberg, Daniel (1961). 'Risk, Ambiguity and the Savage Axioms', *Quarterly Journal of Economics*, **75**, 643–69.

Falk, Barry (1991). 'Formally Testing the Present Value Model of Farmland Prices', *American Journal of Agricultural Economics*, **73**, 1–10.

Fama, Eugene F., and French, Kenneth R. (1988*a*). 'Dividend Yields and Expected Stock Returns', *Journal of Financial Economics*, **22**, 3–25.

—— —— (1988*b*). 'Permanent and Temporary Components of Stock Prices', *Journal of Political Economy*, **96**, 246–73.

—— —— (1992). 'The Cross Section of Expected Stock Returns', *Journal of Finance*, **47**, 427–65.

Feldstein, Martin, and Horioka, Charles (1980). 'Domestic Saving and International Capital Flows', *Economic Journal*, **90**, 314–29.

Fischer, Stanley (1975). 'The Demand for Index-Linked Bonds', *Journal of Political Economy*, **83**, 509–34.

—— (1977). 'Long-Term Contracts, Rational Expectations, and the Optimal Money Supply Rule', *Journal of Political Economy*, **85**, 191–205.

Fischhoff, Baruch, Slovic, Paul, and Lichtenstein, Sara (1980). 'Knowing What You Want: Measuring Labile Values', in Tom Wallsten (ed.), *Cognitive Processes in Choice and Decision Behavior*, Erlbaum, Hillsdale, NJ.

Fisher, Irving (1911). *The Purchasing Power of Money*, Macmillan, London.
—— (1928). *The Money Illusion*, The Adelphi Co., New York.
Folger, Robert, and Martin, Chris (1986). 'Relative Deprivation and Referent Cognitions: Distributive and Procedural Justice Effects', *Journal of Experimental Social Psychology*, **22**, 531–46.
Forsythe, F. F. (1978). 'The Practical Construction of a Chain Price Index', *Journal of the Royal Statistical Society, Series A*, **141** (pt. 3), 348–58.
French, Kenneth R. (1983). 'A Comparison of Futures and Forward Prices', *Journal of Financial Economics*, **12**, 311–42.
—— (1953). *Essays in Positive Economics*, University of Chicago Press.
Friedman, Milton (1962). *Capitalism and Freedom*, University of Chicago Press.
Friend, Irwin, and Blum, Marshall E. (1975). 'The Demand for Risky Assets', *American Economic Review*, **65**, 900–22.
Gammill, James F., Jr., and Perold, Andre F. (1989). 'The Changing Character of Stock Market Liquidity', *Journal of Portfolio Management*, spring, **15**, 13–18.
Garbade, Kenneth, and Silber, William L. (1983). 'Cash Settlement of Futures Contracts: An Economic Analysis', *Journal of Futures Markets*, **3**, 451–72.
Garcia, Marcio G. (1991). 'The Formation of Inflation Expectations in Brazil: A Study of the Futures Market for the Price Level', unpublished working paper, Departamento de Economia, PUC/RJ, Rio de Janeiro.
Gardner, Bruce L. (1989). 'Rollover Hedging and Missing Long-Term Futures Markets', *American Journal of Agricultural Economics*, **71**, 311–18.
Garman, Mark B. (1978). 'The Pricing of Supershares', *Journal of Financial Economics*, **6**, 3–10.
—— (1987). 'Perpetual Currency Options', *International Journal of Forecasting*, **3**, 179–84.
Gehr, Adam (1988). 'Undated Futures Markets', *Journal of Futures Markets*, **88**, 89–97.
Gemmill, Gordon (1990). 'Futures Trading and Finance in the Housing Market', *Journal of Property Finance*, **1**, No. 2, 196–207.
Goetzmann, William N. (1990). 'Accounting for Taste: An Analysis of Art Returns over Three Centuries', unpublished working paper, Columbia University, New York.
—— (1992). 'The Accuracy of Real Estate Indices: Repeat Sale Estimators', *Journal of Real Estate Finance and Economics*, **5**, 5–53.

—— and Spiegel, Matthew (1992). 'Non-Temporal Components of Residential Real Estate Appreciation', unpublished working paper, Columbia University, New York.

Gordon, Robert J. (1992). 'Measuring the Aggregate Price Level: Implications for Economic Performance and Policy', Working Paper No. 3969, National Bureau of Economic Research, Cambridge, Mass.

Gorton, Gary, and Pennacci, George (1991). 'Security Baskets and Index-Linked Securities', Working Paper No. 3711, National Bureau of Economic Research, Cambridge, Mass.

Gray, Roger W. (1977). 'The Futures Market for Maine Potatoes: An Appraisal', in Ann E. Peck (ed.), *Selected Writings on Futures Markets*, Chicago Board of Trade.

Griliches, Zvi (1961). 'Hedonic Price Indices for Automobiles: An Economic Analysis of Quality Change', in Price Statistics Review Committee, *The Price Statistics of the Federal Government*, US Congress, Joint Economic Committee, *Government Price Statistics Hearings*, pt. 1. 87th Congress, 1st Session (also published as National Bureau of Economic Research General Series No. 73, Cambridge, Mass.).

Grossman, Sanford J. (1977). 'The Existence of Futures Markets, Noisy Rational Expectations, and Information Externalities', *Review of Economic Studies*, **44**, 431-49.

Hall, Robert E. (1971). 'The Measurement of Quality Change from Vintage Price Data', in Zvi Griliches (ed.), *Price Indices and Quality Change*, Harvard University Press, Cambridge, Mass.

Harris, Lawrence (1990). 'The Economics of Cash Index Alternatives', *Journal of Futures Markets*, **10**, 179–94.

Haurin, Donald R., Hendershott, Patric H., and Kim, Dongwook (1991). 'Local House Price Indexes: 1982–1991', *AREUEA Journal*, **19**, 451–72.

Heckman, James (1976). 'The Common Structure of Statistical Models of Truncation, Sample Selection, and Limited Dependent Variables and a Simple Estimator for Such Models', *Annals of Economic and Social Measurement*, **5**, 475–92.

Heifner, Richard G., Driscoll, James L., Helmuth, John W., Leath, Mack N., Niernberger, Floyd F., and Wright, Bruce H. (1977). *The US Cash Grain Trade in 1974: Participants, Transactions, and Information Sources*, US Department of Agriculture, Economic Research Service, Agricultural Economic Report No. 386.

Helmuth, John W. (1977). *Grain Pricing*, Commodity Futures Trading Commission, Washington, DC.

Hochman, Shalom, and Palmon, Oded (1988). 'A Tax-Induced Clientele for Index-Linked Bonds', *Journal of Finance*, **43**, 1257–63.

Hogarth, Robin M., and Kunreuther, Howard (1989). 'Risk, Ambiguity and Insurance', *Journal of Risk and Uncertainty*, **2**, 5–35.

Horrigan, Brian R. (1987). 'The CPI Futures Market: The Inflation Hedge that Won't Grow', *Business Review* (Federal Reserve Bank of Philadelphia), May–June.

Hsiao, Cheng (1986). *Analysis of Panel Data*, Cambridge University Press.

Ito, Takatoshi, and Hirono, Keiko Nosse (1993). 'Efficiency of the Tokyo Housing Market', *Bank of Japan Monetary and Economic Studies*, **11**, 1–32.

Jarrow, Robert A., and Oldfield, George S. (1981). 'Forward Contracts and Futures Contracts', *Journal of Financial Economics*, **9**, 373–82.

Johnston, Elizabeth Tashjian, and McConnell, John J. (1989). 'Requiem for a Market: An Analysis of the Rise and Fall of a Financial Futures Market', *Review of Financial Studies*, **2**, 1–23.

Kahneman, Daniel, Knetsch, Jack L., and Thaler, Richard (1986). 'Fairness as a Constraint on Profit Seeking: Entitlements in the Market', *American Economic Review*, **76**, 728–40.

—— and Tversky, Amos (1979). 'Prospect Theory: An Analysis of Decision under Risk', *Econometrica*, **47**, 263–92.

Kallick, M., Suits, D., Dielman, T., and Hybels, J. (1975). *A Survey of American Gambling Attitudes and Behavior*, Survey Research Center, Institute for Social Research, University of Michigan, Ann Arbor.

Katona, George (1975). *Psychological Economics*, Elsevier, New York.

Keynes, John M. (1936). *The General Theory of Employment, Interest and Money*, Macmillan, London.

Kihlstrom, Richard, and Pauly, Mark (1971). 'The Role of Insurance in the Allocation of Risk', *American Economic Review*, **61**, 371–9.

Kim, Sunwoong (1992). 'Search, Hedonic Prices and Housing Demand', *Review of Economics and Statistics*, **74**, 503–8.

Kim, Tae-Dong (1993). 'Asset Price Movements and Bubbles in Korea and Japan' (in Korean), unpublished working paper, Sung Kyun Kwan University, Seoul.

Knetsch, Jack L., Thaler, Richard H., and Kahneman, Daniel (1989). 'Experimental Tests of the Endowment Effect and the Coase Theorem', unpublished working paper, Simon Fraser University.

Knight, Frank (1921). *Risk, Uncertainty and Profit*, Houghton Mifflin, Boston.

Knoll, Michael S. (1991). 'A Tax-Induced Clientele for Index-Linked Bonds: A Comment', *Journal of Finance*, **46**, 1933–6.

Krueger, Alan B., and Bowen, William B. (1993). 'Income-Contingent College Loans', *Journal of Economic Perspectives*, **7**, 193–201.

Kunreuther, Howard (1977). *Disaster Insurance Protection: Public Policy Lessons*, Wiley Interscience, New York.

Kuo, Chiong-Long (1993). 'Serial Correlation and Seasonality in the Real Estate Market', unpublished working paper, Yale University.

Kupiec, Paul (1990). 'A Survey of Exchange-Traded Basket Instruments', *Journal of Financial Services Research*, **4**, 175–90.

Kyle, Albert S. (1984). 'A Theory of Futures Market Manipulation', in Ronald W. Anderson (ed.), *The Industrial Organization of Futures Markets*, Heath, Lexington Books, Lexington, Mass.

Lawley, D. N., and Maxwell, A. E. (1971). *Factor Analysis as a Statistical Method*, American Elsevier, New York.

Lee, Charles, Shleifer, Andrei, and Thaler, Richard (1991). 'Investor Sentiment and the Closed-End Puzzle', *Journal of Finance*, **46**, 75–109.

Lee, Wayne (1975). *Experimental Design and Analysis*, W. H. Freeman, San Francisco.

LeRoy, Stephen F., and Porter, Richard D. (1981). 'Stock Price Volatility: A Test Based on Implied Variance Bounds', *Econometrica*, **49**, 97–113.

Leventhal, Gerald S., Karuza, Jr., Jurgis, and Fry, William Rick (1980). 'Beyond Fairness: A Theory of Allocational Preferences', in G. Mikula (ed.), *Justice and Social Interaction*, Springer, New York.

Lovell, Michael C., and Vogel, Robert C. (1973). 'A CPI-Futures Market', *Journal of Political Economy*, **81**, 1009–12.

Lucas, Robert E. (1978). 'Asset Prices in an Exchange Economy', *Econometrica* **46**, 1429–45.

Maddala, G. S. (1983). *Limited-Dependent and Qualitative Variables in Econometrics*, Cambridge University Press.

Mandel, John (1961). 'Non-Additivity in Two-Way Analysis of Variance', *Journal of the American Statistical Association*, **56**, 878–88.

March, James G. (1978). 'Bounded Rationality, Ambiguity, and the Engineering of Choice', *Bell Journal of Economics*, **9**, 587–608.

Marchitelli, Richard and Peter F. Korpacz (1992). 'Market Value: The Elusive Standard', *Appraisal Journal*, **60**, 313–22.

Mark, J. H., and Goldberg, M. A. (1984). 'Alternative Housing Price Indices: An Evaluation', *AREUEA Journal* **12**, 30–49.

Marshall, John F., Bansal, Vipul, Herbst, Anthony F., and Tucker, Alan L. (1992). 'Hedging Business Cycle Risk with Macro Swaps and Options', *Continental Bank Journal of Applied Corporate Finance*, **4**, 103–8.

McGuire, W. J. (1969). 'The Nature of Attitudes and Attitude Change', in G. Linzey and E. Aronson (eds.), *The Handbook of Social Psychology*, Addison-Wesley, Reading, Mass.

McCulloch, J. Huston (1980). 'The Ban on Indexed Bonds, 1933–77', *American Economic Review*, **70**, 1018–21.

Mehra, Rajnish, and Prescott, Edward C. (1985). 'The Equity Premium: A Puzzle', *Journal of Monetary Economics*, **15**, 145–61.

Merton, Robert C. (1973). 'An Intertemporal Capital Asset Pricing Model', *Econometrica*, **41**, 867–87.
Miles, Mike, Pittman, Robert, Hoesli, Martin, Bhatnagar, Pankaj, and Guilkey, David (1991). 'A First Detailed Look at the Extent of America's Real Estate Wealth', unpublished working paper, University of North Carolina at Chapel Hill.
Miller, Robert (1989). 'Property Price Futures and Options?', *Futures and Options World.*
Montgomery, Claire (1990). *Household Investment in the Improvement of the Existing Housing Stock*, unpublished Ph.D. dissertation, Department of Economics, University of Washington, Seattle.
Morck, Randall, Shleifer, Andrei, and Vishny, Robert W. (1990). 'The Stock Market and Investment: Is the Market a Sideshow?', *Brookings Papers on Economic Activity*, **2**, 157–216.
Noguchi, Yukio (1992*a*). *Bubble Economics* (in Japanese), Nihon Keizai, Tokyo.
—— (1992*b*). 'Japan's Land Problem', *Japanese Economic Studies*, **20**, 51–77.
Obstfeld, Maurice (1993*a*). 'Are Industrial-Country Consumption Risks Globally Diversified?', National Bureau of Economic Research Working Paper No. 4308, Cambridge, Mass.
—— (1993*b*). 'International Capital Mobility in the 1990s', forthcoming in Peter B. Kenen (ed.), *Understanding Interdependence: The Macroeconomics of the Open Economy*, Princeton University Press.
Palmquist, Raymond B. (1979). 'Hedonic Price and Depreciation Indexes for Residential Housing: A Comment', *Journal of Urban Economics*, **6**, 267–71.
—— (1982). 'Measuring Environmental Effects on Property Values without Hedonic Regressions', *Journal of Urban Economics*, **11**, 333–47.
—— (1989). 'Land as a Differentiated Factor of Production: A Hedonic Model and Its Implications for Welfare Measurement', *Land Economics*, **65**, 23–8.
Petzel, Todd E. (1985). 'Consumer Price Index Futures as an Instrument for Hedging Inflation Uncertainty', unpublished working paper, Coffee Sugar and Cocoa Exchange, New York.
—— and Fabozzi, Frank J. (1986). 'Real Interest Rates and CPI-W Futures', *Advances in Futures and Options Research*, **1**, 255–70.
Pierog, Karen, and Stein, Jon (1989). 'New Contracts: What Makes Them Fly or Fail?', *Futures*, **18**, 50–4.
Polanyi, Karl (1944). *The Great Transformation*, Farrar & Rinehart, NY.
Poterba, James M. (1991). 'House Price Dynamics: The Role of Tax Policy and Demography', *Brookings Papers on Economic Activity*, **2**, 143–99.

—— and Summers, Lawrence H. (1988). 'Mean Reversion in Stock Returns: Evidence and Implications', *Journal of Financial Economics*, **22**, 27–59.

Powers, Mark J. (1970). 'Does Futures Trading Reduce Price Fluctuations in the Cash Market?', *American Economic Review*, **60**, 460–4.

Radner, Roy (1968). 'Competitive Equilibrium under Uncertainty', *Econometrica*, **36**, 31–58.

Reich, Robert B. (1992). *The Work of Nations*, Vintage Books, New York.

Rhee, Changyong, and Rhee, Wooheon (1991). 'Fundamental Value and Investment: Micro Data Evidence', Rochester Center for Economic Research Working Paper No. 282, New York.

Romer, Christina D. (1989). 'The Prewar Business Cycle Reconsidered: New Estimates of Gross National Product 1869–1908', *Journal of Political Economy*, **97**, 1–37.

Ross, Stephen (1976). 'The Arbitrage Theory of Capital Asset Pricing', *Journal of Economic Theory* **13**, 341–60.

Rosenzweig, Mark R. (1988). 'Risk, Implicit Contracts, and the Family in Rural Areas of Low Income Countries', *Economic Journal*, **98**, 1148–70.

Rothschild, Michael, and Stiglitz, Joseph (1976). 'Equilibrium in Competitive Insurance Markets: An Essay on the Economics of Imperfect Competition', *Quarterly Journal of Economics*, **90**, 629–49.

Sala-i-Martin, Xavier, and Sachs, Jeffrey (1992). 'Fiscal Federalism and Optimum Currency Areas: Evidence for Europe from the United States', in M. B. Canzoneri *et al.* (eds.), *Establishing a Central Bank: Issues in Europe and Lessons from the U.S.*, Cambridge University Press.

Shiller, Robert J. (1981). 'Do Stock Prices Move Too Much to be Justified by Subsequent Changes in Dividends?', *American Economic Review*, **71**, 421–36.

—— (1989). *Market Volatility*, MIT Press, Cambridge, Mass.

—— (1990). 'Market Volatility and Investor Behavior,' *American Economic Review*, **80**, 58–62.

—— (1991). 'Arithmetic Repeat Sales Price Estimators', *Journal of Housing Economics*, **1**, 110–26.

—— (1992). 'Who's Minding the Store?', in *The Report of the Twentieth Century Fund Task Force on Market Speculation and Corporate Governance*, The Twentieth Century Fund Press, New York.

—— (1993*a*). 'Aggregate Income Risk and Hedging Mechanisms', National Bureau of Economic Research Working Paper, Cambridge, Mass.

—— (1993*b*). 'Measuring Asset Values for Cash Settlement in Derivative Markets: Hedonic Repeated Measures Indices and Perpetual Futures', *Journal of Finance*, July.

Shlaes, Jared (1984). 'The Market in Market Value', *Appraisal Journal*, **52**, 494-518.
Siegel, Jeremy J. (1994). *Stocks for the Long Run*, Irwin, New York.
Slovic, Paul, and Lichtenstein, Sarah (1983). 'Preference Reversals: A Broader Perspective', *American Economic Review*, **73**, 596–605.
Stein, Jerome L. (1986). *The Economics of Futures Markets*, Basil Blackwell, Oxford, England.
Stevens, Neil A. (1974). 'The Futures Market for Farm Commodities: What It Can Mean to Farmers', *Federal Reserve Bank of St. Louis Review*, **56**, 10–5.
Stockman, A., and Tesar, L. (1990). 'Tastes and Technology in a Two Country Model of the Business Cycle: Explaining International Comovements', Working Paper No. 3566, National Bureau of Economic Research, Cambridge, Mass.
Summers, Robert, and Heston, Alan (1988). 'A New Set of International Comparisons of Real Product and Price Levels: Estimates for 130 Countries', *Review of Income and Wealth*, **34**, 1–25.
—— —— (1991). 'The Penn World Table (Mark 5): An Expanded Set of International Comparisons', *Quarterly Journal of Economics*, **106**, 1–41.
Taylor, John B. (1980). 'Aggregate Dynamics and Staggered Contracts', *Journal of Political Economy*, **88**, 1–23.
Thaler, Richard (1991). *Quasi Rational Economics*, Russell Sage Foundation, New York.
Tirole, Jean (1982). 'On the Possibility of Speculation under Rational Expectations', *Econometrica*, **50**, 1163–81.
Tobin, James (1973). 'Financing Higher Education', unpublished working paper, Yale University, New Haven, Conn.
Townsend, Robert M. (1993). 'Risk and Insurance in Village India', *Econometrica*, forthcoming.
Triplett, Jack E. (1983). *The Measurement of Labor Costs*, University of Chicago Press.
—— (1990). 'Hedonic Methods in Statistical Agency Environments: An Intellectual Biopsy', in Ernst R. Berndt and Jack E. Triplett (eds.), *Fifty Years of Economic Measurement*, National Bureau of Economic Research and University of Chicago Press.
Tufano, Peter (1992). 'Financial Innovation and First Mover Advantages', *Journal of Applied Corporate Finance*, **5**, 83–7.
Turvey, Calum G., and Baker, Timothy G. (1990). 'A Farm Level Financial Analysis of Farmers' Use of Futures and Options under Alternative Farm Programs', *American Journal of Agricultural Economics*, **72**, 946–57.
Tversky, Amos, and Kahneman, Daniel (1981). 'The Framing of Decisions and the Psychology of Choice', *Science*, **211**, 453–8.

Tyler, Tom, and Caine, Andrew (1981). 'The Influence of Outcomes and Procedures on Satisfaction with Formal Leaders', *Journal of Personality and Social Psychology*, **41**, 642–55.
US Department of Commerce, Bureau of Economic Analysis (1973). *Long Term Economic Growth 1860–1970*, US Government Printing Office, Washington, DC.
—— (1992). *National Income and Product Accounts of the United States, Volume 2, 1959–88*, US Government Printing Office, Washington, DC.
US Department of Labor, Bureau of Labor Statistics (1986). *Producer Price Measurement: Concepts and Methods*, Washington, DC.
—— (1992). *BLS Handbook of Methods*, Bulletin 2414, US Government Printing Office, Washington, DC.
US Panel on Educational Innovation, President's Science Advisory Committee (Jerrold R. Zacharias Panel) (1967). *Educational Opportunity Bank*, US Government Printing Office, Washington, DC.
US Securities and Exchange Commission, Division of Market Regulation (1988). *The October 1987 Market Break*, US Government Printing Office, Washington, DC.
Webb, Brian, Miles, Mike, and Guilkey, David (1993). 'Transactions-Driven Commercial Real Estate Returns: The Panacea to Asset Allocation Models?', *Journal of the American Real Estate and Urban Economics Association*, **20**, 325–57.
Webb, Cary (1988). 'A Probabilistic Model for Price Levels in Discontinuous Markets,' in W. Eichhorn (ed.), *Measurement in Economics*, Physica, Heidelberg.
Weinstein, Neil D. (1989*a*). 'Effects of Personal Experience on Self-Protective Behavior', *Psychological Bulletin*, **105**, 31–50.
—— (1989*b*). 'Optimistic Biases about Personal Risks', *Science*, **246**, 1232–3.
Welch, Ivo (1990). 'Sequential Sales, Learning, and Cascades', unpublished working paper, University of California, Los Angeles.
West, Kenneth D. (1988*a*). 'Bubbles, Fads, and Stock Price Volatility: A Partial Evaluation', *Journal of Finance*, **43**, 639–55.
—— (1988b). 'Dividend Innovations and Stock Price Volatility', *Econometrica*, **56**, 37–61.
Williamson, Oliver (1979). 'Transaction-Cost Economics: The Governance of Contractual Relations', *Journal of Law and Economics*, **22**, 233–61.
Wold, Hermann (1966). 'Nonlinear Estimation by Iterative Least Square Procedures' in F. N. David (ed.), *Research Papers in Statistics*, Wiley, London.
Working, Holbrook (1953). 'Futures Trading and Hedging', *American Economic Review*, 314–43.

Author Index

Subject Index

9 780198 294184

Printed in the United States
By Bookmasters

Printed in the United States
By Bookmasters